AF361926

Wireless Rechargeable Sensor Networks 2019

Wireless Rechargeable Sensor Networks 2019

Editor

Chang Wu Yu

MDPI • Basel • Beijing • Wuhan • Barcelona • Belgrade • Manchester • Tokyo • Cluj • Tianjin

Editor
Chang Wu Yu
Department of Computer Science
and Information Engineering,
Chung Hua University
Taiwan

Editorial Office
MDPI
St. Alban-Anlage 66
4052 Basel, Switzerland

This is a reprint of articles from the Special Issue published online in the open access journal *Energies* (ISSN 1996-1073) (available at: https://www.mdpi.com/journal/energies/special_issues/WRSN_2019).

For citation purposes, cite each article independently as indicated on the article page online and as indicated below:

LastName, A.A.; LastName, B.B.; LastName, C.C. Article Title. *Journal Name* **Year**, *Article Number*, Page Range.

ISBN 978-3-03943-599-9 (Hbk)
ISBN 978-3-03943-600-2 (PDF)

Contents

About the Editor

Chang Wu Yu (Professor) was born in Taoyuan, Taiwan, in 1964. He received his BS degree from Soochow University in 1985, MS degree from National Tsing Hua University in 1989, and Ph.D. degree from National Taiwan University in 1993, all in Computer Sciences. Currently, he is Professor at the Department of Computer Science and Information Engineering, Chung Hua University, Taiwan. Prof. Yu has published more than 80 papers including in *IEEE Journal on Selected Areas in Communications*, *IEEE Transactions on Parallel and Distributed Systems*, *Theoretical Computer Science*, ACM/Springer *Wireless Networks*, *ACM Transactions on Embedded Systems*, and Elsevier's *Ad Hoc Networks*. Dr. Yu received the Best Paper Award at the 2008 ACM International Conference on Sensors, Ad Hoc, and Mesh Networks at the 9th Workshop on Wireless, Ad Hoc, and Sensor Networks (WASN 2013), and at both the 2004 and 2007 Mobile Computing Workshop. Dr. Yu serves on the editorial boards of number international journals including *Ad Hoc & Sensor Wireless Networks*. Moreover, Dr. Yu has serves as Guest Editor for several international journals. His current research interests include fundamental theories, algorithms, and applications in wireless networks, wireless sensor networks, and distributed computing by exploiting graph techniques.

Preface to "Wireless Rechargeable Sensor Networks 2019"

Wireless sensor networks have attracted much attention recently due to their various applications in many fields. Due to limited power consumption, these sensor nodes may experience power shortages and, thus, lead to many problems including network disconnection. Most previous methods have focused on providing energy-saving strategies to elevate the lifetime of sensor networks. Another aggressive but different approach is to wirelessly recharge sensor nodes to increase the lifetime of the sensor networks. This Special Issue, entitled "Wireless Rechargeable Sensor Networks", invited articles that address state-of-the-art technologies and new developments for wireless rechargeable sensor networks (WRSNs). Particularly encouraged was the submission of articles dealing with the latest hot topics in WRSNs, such as charger deployment, charger scheduling, wireless energy transfer, mobile charger design, energy-harvesting technique, and energy provisioning. Of interest as well were articles which discuss protocols, algorithms, and optimization in WRSN.

For this Special Issue, we received many submissions from different countries all around the globe in response to our call for papers. Each accepted article has been reviewed by at least three reviewers. We believe that the accepted papers present the most up-to-date progress in algorithms and theory for robust wireless sensor networks with respect to different networking problems. First of all, we would like to thank our authors, who provided their remarkable contributions to this Special Issue. We also appreciate the outstanding review work performed by the referees of this Special Issue who provided valuable comments to the authors. Without their full support, this Special Issue would not have been such a success.

Chang Wu Yu
Editor

Article

A Novel Hybrid Search and Remove Strategy for Power Balance Wireless Charger Deployment in Wireless Rechargeable Sensor Networks

Tu-Liang Lin, Hong-Yi Chang * and Yu-Hsin Wang

Department of Management Information System, National Chiayi University, Chiayi 60054, Taiwan;
tuliang@mail.ncyu.edu.tw (T.-L.L.); s1064629@mail.ncyu.edu.tw (Y.-H.W.)
* Correspondence: hychang@mail.ncyu.edu.tw

Received: 29 February 2020; Accepted: 20 May 2020; Published: 25 May 2020

Abstract: Conventional sensor nodes are often battery-powered, and battery power limits the overall lifetime of the wireless sensor networks (WSNs). Wireless charging technology can be implemented in WSNs to supply power to sensor nodes and resolve the problem of restricted battery power. This type of mixed network is called wireless rechargeable sensor networks (WRSNs). Therefore, wireless charger deployment is a crucial task in WRSNs. In this study, the method of placing wireless chargers to efficiently extend the lifetime of the WRSNs is addressed. Owing to the data forwarding effect in WSNs, sensor nodes that are closer to the data collection or sink node drain more power than nodes that are further away from the data collection or sink node. Therefore, this study proposes a novel hybrid search and removal strategy for the power balance charger deployment method. The wireless chargers are placed in the chosen nodes of the WRSNs. The node-chosen problem we address is called the dominating set problem. The proposed hybrid search and removal strategy attempts to discover the minimum number of chargers required to cover all sensor nodes in the WRSN. The proposed algorithm considers the charging power of the wireless directional charger when arranging its placement to maximize the charging capacity in a power-balanced prerequisite. Therefore, the proposed deployment strategy preserves the awareness of the presence of the sink node that could result in unbalanced power distribution in WRSNs. The simulation results show that the proposed strategy spares more chargers and achieves better energy efficiency than other deployment approaches.

Keywords: wireless rechargeable sensor networks; hybrid method; wireless charger deployment

---

1. Introduction

In a wireless sensor network (WSN), the amount of power provided by the battery determines whether each sensor can operate normally. Such power limitations also affect the stability of the entire WSN. To solve the problem of limited battery power, scholars have proposed studies on the design of energy-efficient routing protocols [1] and energy-harvesting wireless sensor networks (WSNs) to reduce energy consumption [2]. Neither the energy-efficient protocols nor the energy-harvesting WSNs solves the principle problem. These studies can prolong the life of WSNs. However, when the sensor exhausts the battery, the WSNs still cannot work properly. Some might argue that WSNs using solar energy can solve the problem of battery exhaustion. However, solar energy-based sensors are affected by environmental factors, and the efficiency of WSNs is significantly affected by different environments. In recent years, some researchers have proposed wireless charging technology, which can solve the problem of sensor battery power limitation, thus fundamentally solving the WSN lifetime problem [3]. The power can be transmitted using wireless chargers to RFIDs [4], sensors [5] and other terminal devices. TechNavio's market research shows that wireless charging technology has great market potential, and the 2020 global annual increase in the wireless charging market is more than 33% [6]. Due to the convenience and continuous development of wireless charging technology, wireless charging has drawn the attention of both industry and academia, especially in the realm of WSN research.

The adoption of wireless charging technology to improve the survival time of WSNs has been extensively surveyed. The new type of WSN that allows wireless chargers to extend the lifetime of wireless sensors is called wireless rechargeable sensor networks (WRSNs). The main consideration for the deployment of wireless chargers in WSNs is to maintain a stable power supply for the sensors so that the sensors can perform sensing tasks in the sensing area. Based on cost considerations, the number of wireless chargers in this deployment area should be minimized. However, because the power supply of the sensors has a significant impact on the performance of the WSNs, this study attempts to find a deployment strategy that can effectively reduce the number of wireless chargers while maximizing the charging capacity.

Several studies related to the deployment of wireless chargers in WRSNs have been proposed in recent years. However, these studies have some limitations, such as the grid partition that does not fit to the real environment in which the charger can be placed. Some might unrealistically assume that the data collection node or sink node does not exist. The data collection node or sink node in WSNs or WRSNs is a critical device. The sensors collect and route data to a data collection or sink node. If a sensor is near the data collection or sink node, it will bear frequent data forwarding and consume more energy. Therefore, to address the limitations of previous research, this study explores the deployment of directional chargers in a given network. In this study, the influence of data collection or sink nodes is evaluated. Previous studies have failed to address this influence. We also deal with two different types of chargers: omnidirectional and directional chargers in WRSNs. In this study, a two-stage strategy called search most and remove useless (SMRU) is proposed. In the first stage, the candidate position of the directional charger is selected on a sensor randomly. Next, we search for the number of sensors covered by the chargeable radius at the candidate position of the directional charger. The deployment given by the proposed SMRU strategy can effectively reduce the required number of directional chargers and preserve good charging utility. The experimental results demonstrate that our proposed SMRU approach outperforms previous works.

The three main contributions of the proposed SMRU algorithm are as follows:

1. The SMRU algorithm can deploy the chargers in the WRSN in less time, and the number of chargers required is relatively small.
2. When the sensor position moves, the SMRU algorithm takes less time to correct and deploy the charger positions.
3. The SMRU algorithm can calculate the energy consumption of the sink node. Under the premise of supplying enough power to the sink node, the mechanism of the SMRU algorithm can reduce the number of chargers used to charge the sink node without affecting the number of chargers that charge other sensors.

The structure of this study is as follows. In Section 2, this study investigates the relevant literature and explores the problems of some deployment strategies. In Section 3, the problem description is presented. Section 4 presents the details of the proposed algorithms. Section 5 explains the experimental results and give conclusions.

2. Related Works

2.1. Omnidirectional and Directional Chargers

Previous studies assumed that omnidirectional chargers have the same charging capacity at the same distance [7,8]. When the distance increases, the received power decays, and when the threshold distance is finally reached, the received power becomes zero. The charging range of an omnidirectional charger is defined as the center of the charging range. The coverage area of the directional charger is a sector [7,8]. Devices from different manufacturers may have different ranges of charging angles. Due to the somewhat steady power coverage and low cost, directional chargers are frequently used in WRSNs.

As the mathematical formulation of a charger's charging power, previous researchers defined the charging formula as follows (1) [9–11]:

$$W(dist(a_i, s_i)) = \begin{cases} \frac{\alpha}{(dist(a_i,s_i)+\beta))}, & dist(a_i, s_i) \leq r \\ 0, & dist(a_i, s_i) > r \end{cases}. \tag{1}$$

In Equation (1), $dist(a_i,s_i)$ is the Euclidean distance between charger a_i and sensor s_i $\alpha = \frac{\xi W_0 G_a G_s}{C_W}\left(\frac{\lambda}{4\pi}\right)^2$, λ is the wavelength, and ξ is the utility of the rectifier. G_a and G_s are the antenna gains of the charger and the sensor, respectively. W_0 is the charger power and β is the compensation parameter for short-distance transmission and C_w in α is polarization mismatch loss [9]. When the range is larger than the charging radius r, the charging power is set to 0.

2.2. Wireless Charger and Sensor Deployment

The deployment of wireless chargers has drawn much attention recently. Several studies have been conducted in this area. He et al. proposed an energy provisioning method for wireless rechargeable sensor networks [12]. The proposed method addressed omnidirectional wireless charger deployment, and the wireless chargers were deployed in the region of interest to supply sufficient charging power to support the operations in WRSNs. However, He et al. probe the region using a conventional triangular method. Chiu et al. proposed a mobility-aware charger deployment scheme for WRSNs [13]. The wireless charger placement methods proposed by Chiu et al. [13] considered the trajectories of mobile sensors. The methods partitioned the regions into grids, and omnidirectional wireless chargers were arranged on grid intersections. Liao et al. scheduled the sleep period of wireless chargers to reduce the wastage of battery power and adopted a conical structure to cover the region [7]. In contrast to studies on omnidirectional sensor deployment, some researchers have carried out studies on directional sensor deployment, which present the directional angle concept [8,14]. Horster and Lienhart discussed the placement of visual sensors that are directional, and the angles of the devices are considered [8]. According to Horster and Lienhart, the monitoring area of a visual sensor is directional and triangles are used to represent the scope [8]. Therefore, they proposed the integer linear programming solving technique to calculate the optimal coverage, and every grid intersection represented a possible site of the placement [8]. As the deployment of wireless chargers can be formulated as a nonlinear programming problem, the deployment problem befits to the NP hard problem class [15]. Therefore, solving the wireless charger deployment problem is a challenging task. Han et al. studied the deployment of directional sensors [14]. In their study, the number of deployed directional sensors was minimized, and the connections among the sensors were considered. Mo et al. addressed the coordination problem among multiple mobile chargers and formulated the multiple mobile charger coordination problem as mixed-integer linear programming and proposed a decomposition approach to solve the problem [16]. Tang et al. simultaneously considered both charging and routing and proposed an optimization approach to extend the lifetime of the network [17]. In order to balance the network energy, charging efficiency is dynamically balanced and the charging time is partitioned according to the energy consumption. Chen et al. proposed a WRSN model equipped charging drone [18]. A scheduling algorithm to solve the shortest multi-hop path is developed for the charging drone.

Most studies did not address the impact of unbalanced battery power distribution caused by data forwarding to the sink node. Unbalanced battery power distribution can make some sensor nodes, especially those near the sink node, lose power quicker than other nodes. Previous research did not consider the existence of a data collection node or sink node. Although the assumption can allow chargers in an actual environment to be better optimized for deployments, the deployments would actually result in an unbalanced battery power consumption issue. To address the unbalanced battery power distribution controversy, Lin et al. proposed a power balance aware deployment (PBAD) method for wireless chargers and compared the PBAD with random position and random orientation (RPRO) [10]. They formulated the deployment as a minimum dominating set problem and attempted to discover the minimal number of chargers required to cover the networks in the WRSNs. Lin et al. adopted a greedy algorithm to calculate the coverage set, and the influence of the data sink node was weighted to attain full coverage [10]. Lin et al. proposed a two-stage method. First, the chosen region was split into several sub-areas, so that each sub-area could have a continuous supply of charging power. Next, every sub-area was further evaluated to find the minimum dominating sets, after which an approximated optimal dominating set was chosen. The reviewed literature is tabulated in Table 1.

Table 1. Related deployment literature.

Research	Sensor or Charger	Charging Type	Important Concepts		Balanced Battery Power Distribution
He et al., (2012) [12]	Charger	Omnidirectional	1.	Energy provisioning in WRSNs	No
			2.	Physical characteristics of wireless charging	
Chiu et al., (2012) [13]	Charger	Omnidirectional	1.	Mobility-Aware charger deployment	No
			2.	Optimization of mobile deployment	
Liao et al., (2013) [7]	Charger	Omnidirectional	1.	Optimized charger deployment for WRSNs	No
			2.	Sleep scheduling	
Horster and Lienhart., (2006) [8]	Sensor	Directional	1.	Approximating optimal visual sensor placement	No
			2.	Grid linear programming	
Han et al., (2008) [14]	Sensor	Directional	1.	Deploying directional sensor networks	No
			2.	Guaranteed connectivity and coverage	
Lin et al., (2016) [10]	Charger	Directional and Omnidirectional	1.	Power balance aware wireless charger deployment	Yes
			2.	Minimum dominating set problem	
Mo et al. (2019) [16]	Charger	Omnidirectional	1.	The coordination problem among multiple mobile chargers	No
			2.	Mixed-integer linear programming problem	
Tang et al., (2020) [17]	Charger	Omnidirectional	1.	Considering both charging and routing	Yes
			2.	Partitioned charging time according to the energy consumption	
Chen et al., (2020) [18]	Charger	Drone	1.	A WRSN model equipped charging drone	No
			2.	The shortest multi-hop path problem	

3. Search Most and Remove Useless Algorithm

3.1. Problem Definition

There are several problems with wireless sensor network applications. Among them, the power and cost of sensors are often the focus of research. In a wireless sensing network, the sensor must operate continually to maintain the function of receiving and transmitting the information. Therefore, reducing the sensor's power consumption and extending battery life is an important problem. The network architecture called wireless rechargeable sensor network (WRSN) relies on the chargers to continuously supply the sensors with continuous power. In far-field phased arrays, wireless power can be transmitted through beams, while power beam transmission technology can transmit energy over longer distances. The wireless power transfer method proposed in the simulation design in this research is far-field phased array antennas [19]. As the deployment of wireless sensing networks may use a large number of sensors, the use of fewer chargers in WRSNs to provide sufficient power for the sensors to reduce deployment costs and maintain the operation of WRSNs is an important issue. Therefore, this study proposes a method that uses fewer directional chargers to cover all sensors in WRSNs. In our proposed method, the deployment position of the charger can be calculated in a short time. Additionally, all sensors in WRSNs are covered with a small number of chargers, and all sensor power is supplemented to avoid sensor interruption due to power consumption causing WRSNs to fail, and to reduce the cost of deploying a WRSN.

3.2. The SMRU Algorithm

The algorithm is divided into two parts. In the first part, the candidate position of the directional charger is selected on a sensor randomly. Next, we search for the number of sensors covered by the chargeable radius at the candidate position of the directional charger. The candidate position rotates 360° to find the angle that can cover most of the sensors within the charging range. If the number of sensors covered by the candidate position of the directional charger is more than three, the algorithm is repeated for each candidate position within the chargeable radius. To find the angle that can cover most of the sensors within the charging range, the position that covers the most sensors is selected as the new position of the directional charger. If the number of sensors covered by the candidate position of the directional charger is less than three, the candidate position is determined as the position of the directional charger. In the second part, after determining the position of the directional charger, the proposed algorithm checks whether there is an unnecessary position of the directional charger. Therefore, the algorithm checks and removes the charger, and the sensors covered by the charger can be covered by other chargers. This check action can effectively reduce the number of chargers. The operation of SMRU represented as Algorithm 1 is as follows:

Algorithm 1 Search the least needs of the chargers

Input: The locations of all the sensors s_i and directional charger c_i
Output: Positions of chargers
1: **Step 1.** Randomly select the sensor position as the position of the charger.
2: **Step 2.** Search for the number of sensors covered by the chargeable radius d at charger position n.
3: **while** $(distance(s_i - c_i) \leq d)$ **do**
4:     Record the number of sensors to n in the range.
5: **end while**
6: **Step 3.** The charger rotates 360° to find the angle that can cover most of the sensors within the charging range.
7: **Step 4.** If the number of sensors covered by Step 2 is more than three, Step 3 is repeated for different positions of sensors within the chargeable radius, and the position that covers most of the sensors is selected as the new position of the charger. If the number of sensors covered by Step 2 is less than three, the position of the charger does not change.

8: **if** $(n \geq 3)$ **then**
9:     Search 360° and find the best angle that can cover most of the sensors;
10:     Record the number of covered sensors;
11:     Compare to the number of covered sensors of other chargers;
12:     Record the best charger location and the covered sensors in the range;
13: **else**
14:     Search 360° and find the best angle that can cover most of the sensors;
15:     Record the best charger location and the covered sensors.
16: **end if**
17: **Step 5.** Repeat Step 2 until all sensors are covered.
18: **Step 6.** Remove the excess charger positions in which the charger and the sensors covered by the charger can be covered by other chargers.
19: Search all locations of chargers and their covered sensors;
20: **if** (the charger location is redundant) **then**
21:     Remove the redundant charger location;
22: **end if**
23: **return**

3.3. An Example of SMRU

In this study, we use the sensor position to deploy a charger, which can ensure that the worst result of the position selection will converge to the number of sensors. This means that when there are N sensors in WRSNs, it is only necessary to deploy N chargers in the worst case. This can prevent the position selection algorithm from converging. The first idea at the beginning of this study is to search for the charger position that first covers the most sensors among all sensor positions. However, this method is time-consuming because the same calculation is repeated at 360° angles of the charger for each position. Therefore, this study improves the first idea. This study found that the key point to covering all sensors in WRSNs with fewer chargers is the charger position with dense sensors. For example, under the same sensor position configuration, if the charger position is selected as shown in Figure 1, the charge angle of the charger is limited, so that two chargers are required to fully cover all the sensors in this area. However, if the algorithm first compares the positions in the dense area and finds the position that can cover most of the sensors, the algorithm can reduce the number of chargers to one, as shown in Figure 2.

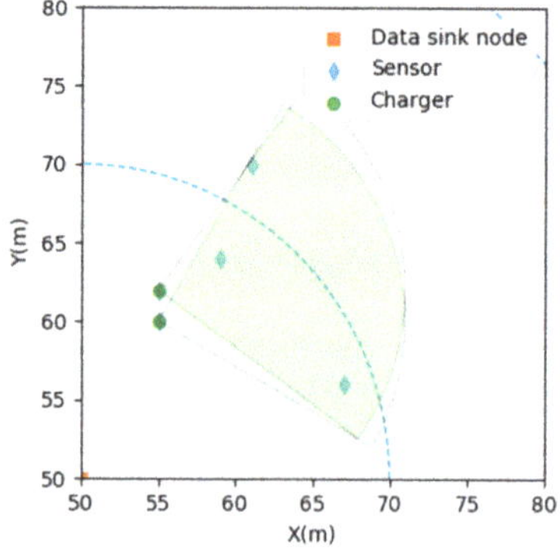

Figure 1. Covers all sensors with two chargers.

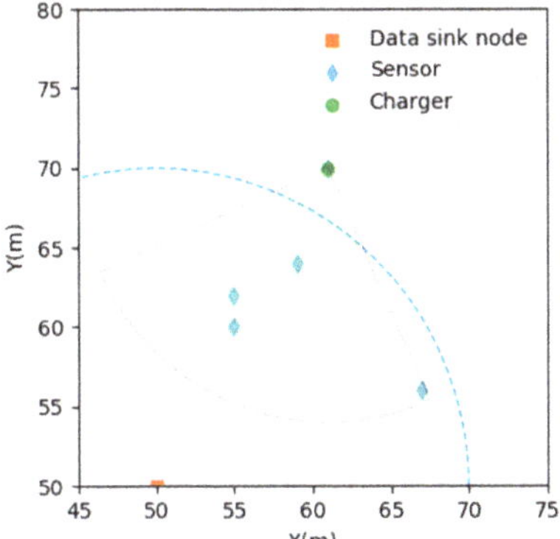

Figure 2. Covers all sensors with one charger.

Therefore, the proposed algorithm first randomly assigns the position of the charger and then checks to see if there are more than three sensors in the 360° charging range of the charger, after which it selects the charger position that covers most of the sensors. If not, then directly select this position and find the best charging angle of the position to save the waste of repeated calculation time caused by the first idea. The reason for choosing more than three sensors in the charging range as the comparison standard is that if there are only one or two sensors that can be covered in the charging range, regardless of the selected position, the number of chargers in the charging range will not be affected. However, if the number of sensors that can be covered in the charging range is more than three, the results may be affected. As shown in Figure 3, if the algorithm selects the middle position, all sensors require two chargers. However, if the algorithm selects the two surrounding positions, it only needs one charger, as shown in Figure 4.

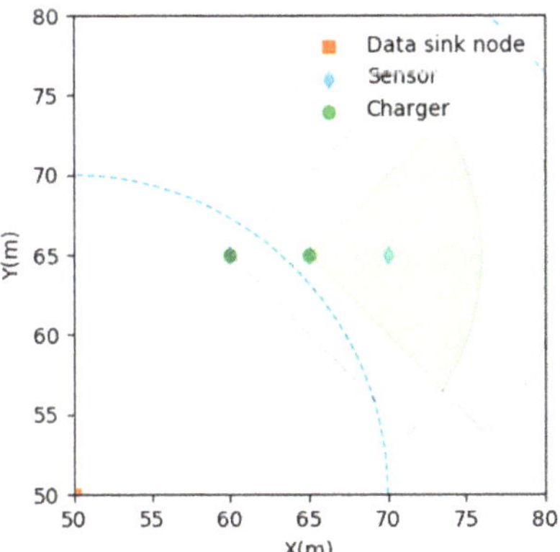

Figure 3. Covers all sensors with two chargers.

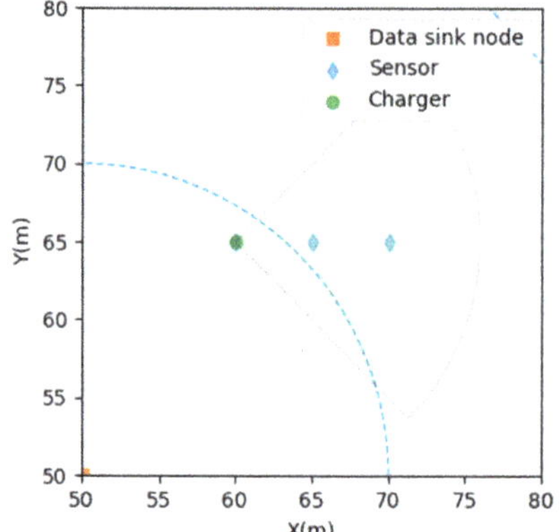

Figure 4. Covers all sensors with one charger.

Finally, although the proposed algorithm minimizes the number of chargers required, experimental results show that extra chargers are required in some cases. The main reasons are the random selection of positions as a sequence of charger deployment, the comparison of the other candidate positions in the charging coverage area of the charger, and the selection of the best position from the candidate positions. As shown in Figure 5, if there are sensors that are arranged continuously, but the distance exceeds the chargeable radius, it cannot be applied to the proposed algorithm. Assume that there are four consecutive sensor positions arranged horizontally. If the order of the positions of the chargers is not appropriate, three chargers are required to fully cover all the sensors in this area. Therefore, this study improves the proposed algorithm again. After the initial selection of the charger position, it is necessary to check whether each charger and the sensors covered by each charger can be covered by other chargers. Assuming that the position of the charger fits this context, it means that the position of this charger is redundant. Therefore, after checking, remove the excess charger positions, as shown in Figure 6.

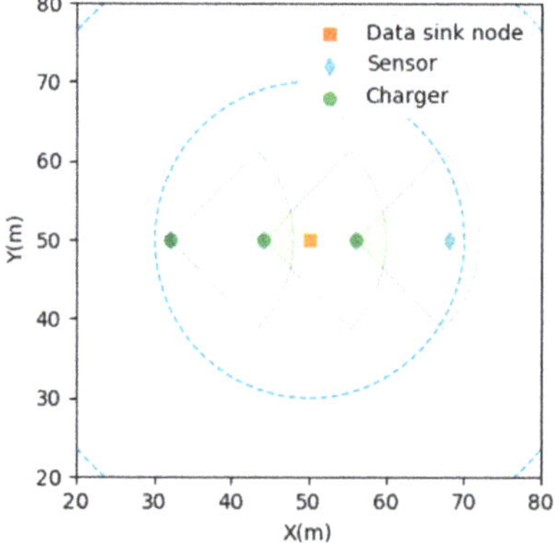

Figure 5. The position of this charger is redundant.

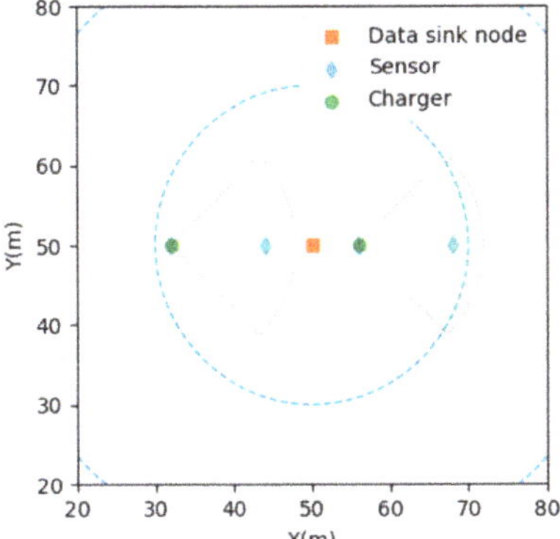

Figure 6. Remove the excess charger position.

3.4. Sensor Mobility

In this study, we consider how the SMRU algorithm can change the position of chargers when the position of the sensors is moved. After the sensor moves the SMRU algorithm first confirms whether the number of sensors covered by the charger has changed. If the number of covered sensors decreases, the charger is rotated 360° to find the charging angle that can cover most of the sensors. If the number of covered sensors is unchanged or increased, the sensor data will be updated. Next, the SMRU algorithm examines whether there are any sensors in the WRSNs that are not being charged by any charger. If it is found that there are uncharged sensors, increase the chargers at the position of this sensor, and then rotate 360° to find the best angle that covers the largest number of sensors. Finally, the SMRU algorithm performs an optimization operation. If it finds an excess charger, it cancels the deployment at that position. The sensor movement was not considered in the PBAD algorithm. As PBAD does not have a mechanism for sensor movement, it needs to be recalculated when the sensors are moved, which takes twice as much time. However, the SMRU algorithm only needs to modify the position of chargers that are affected by sensor movement. Therefore, the modified execution time is much shorter than the original SMRU execution time. The algorithm is shown in Algorithm 2.

Algorithm 2 Changing the charger position after sensor mobility

---

Input: New locations of all sensors and original locations of chargers.
Output: New locations of all chargers.
1: **Step 1.** Check the number of sensors covered by each charger position n.
2: **Step 2.** If the number of covered sensors decreases, the charger is rotated 360° to find the charging angle that can cover most of the sensors.
3: **if** ($n <$ original covered number of sensors) **then**
4:      Search 360° and find the best angle that can cover most of the sensors;
5:      Record the best charger location and the covered sensors;
6: **else**
7:      Record the covered sensors;
8: **end if**
9: **Step 3.** Check if any sensor is uncharged.
10: **Step 4.** If it is found that there are sensors being uncharged, increase the chargers at the position of this sensor, and then rotate 360° to find the best angle that covers the largest number of sensors.
11: **if** (the sensor is not charged) **then**
12:      Add a new charger on the location of the sensor;

13:     Search 360° and find the best angle that can cover most of the sensors;
14:     Record the best charger location and the covered sensors;
15: **end if**
16: **Step 5.** Remove the excess chargers
17: **if** (the charger location is redundant) **then**
18:     Remove the redundant charger location;
19: **end if**
20: **return**

3.5. The Charging of Sink Nodes

The SMRU algorithm can calculate the energy consumption of the sink nodes. Under the premise of supplying enough energy to the sink nodes, the SMRU algorithm can reduce the number of chargers deployed to charge sink nodes. It also does not affect the number of chargers that need to charge sensors. First, the sensor will transfer the data to the sink that is closest to itself. After the SMRU has deployed the charger, it checks the amount of energy consumed by each sink. Next, the SMRU checks the charger within the charging distance of the sink and checks the number of sensors charged by the charger. If the number is equal to one, the charging angle is turned to sink. Therefore, the charger will charge one sensor and one sink. If the number of charging sensors is greater than 1, the charger rotates 360° again to search for the best angle, and the calculation includes the sink node. If the charging target can be covered more than the previous number (meaning that the previously charged sensors are covered; it also covers the sink node), the charging angle is updated; otherwise, the charger maintains the original charging angle. Finally, according to the energy that the sink still lacks, the SMRU deploys additional chargers around the sink to charge the sink. The number of chargers deployed can supplement enough energy as the minimum requirement. Consequently, all sink nodes can have enough power to receive the data sent by the sensor. The SMRU can make good use of the existing charger resources to charge the sink, so it can reduce the number of additional charger deployments. The algorithm is shown in Algorithm 3.

Algorithm 3 Charging energy of the sink nodes

Input: Locations of all sensors, locations of all chargers, and locations of all sinks
Output: Locations of new chargers
1: **Step 1** Check how much energy is consumed by each sink.
2: **Step 2.** Find the charger, and if its distance from the sink node is less than the charging distance r.
3: **Step 3.** If the number of sensors n covered by the charger is greater than one, rotate 360° to check whether the charging of the sink node can be increased without affecting the originally charged sensors. If possible, update the angle of the charger. If not, maintain the original charging angle of the charger. If the number of sensors n covered by the charger is equal to 1, then the charging angle of the charger is turned to the sink node.
4: **if** $(n > 1)$ **then**
5:     Search 360° and find the best angle that can cover most of the sensors;
6:     Check whether the new covered number (ncn) is bigger than the original covered number (ocn);
7:     **if** (ncn > ocn) **then**
8:         Record the new charging angle;
9:     **end if**
10: **else if** $(n == 1)$ **then**
11:     Update charging angle;
12: **end if**
13: **Step 4.** Check the lack of energy for each sink node. Around the sink nodes, randomly select the position of the chargers to deploy the number of chargers needed to supplement the energy required by the sink nodes.
14: **return**

---

4. Experiments and Results

4.1. Experimental Environment and Parameters

In this study, a directional charger was used to conduct a simulation experiment. The experimental environment was to randomly generate sensor positions on a plane with a length of 100 units and a width of 100 units. The proposed algorithm was repeated 500 times to find the average result, and obtain the number of chargers and average execution time. As this study uses a directional charger, its related parameters are set to a chargeable angle of 90° and a chargeable radius of 16 units. Assume that the reception energy consumption of the sink node is 0.6 units/reception. The performance of the proposed algorithm is compared with that of the random position and random orientation (RPRO) and power balance aware deployment (PBAD) [10] algorithms using directional chargers.

4.2. Experimental Results of the Number of Chargers

This experiment compares the relationship between the number of sensors, different algorithms, and the number of chargers required. The average number of chargers used by the SMRU algorithm is less than that used by RPRO and PBAD. Figure 7 shows the average number of chargers used. RPRO selects the charger position and the charging direction randomly. As the charging angle is not selected, the number of chargers required is significantly large. PBAD searches for 360° angles that can cover most of the sensors for charging after randomly selecting positions. Additionally, PBAD repeatedly calculates the same sensor position map and selects the best position using a minimal charger as the deployment location of the charger. In this experiment, PBAD was repeated 10 times to select the best result. Therefore, the experimental results of PBAD are superior to those of RPRO. SMRU first selects a random position and then searches for 360° of the charging radius at that position. If there are more than three sensors (three candidate positions) within the charging radius at this position, it chooses the candidate position that can cover the largest number of sensors and deploy the charger. Additionally, the elimination of the excess charger positions that may be generated in the previous steps means that the number of charger requirements can be reduced to less than that of PBAD.

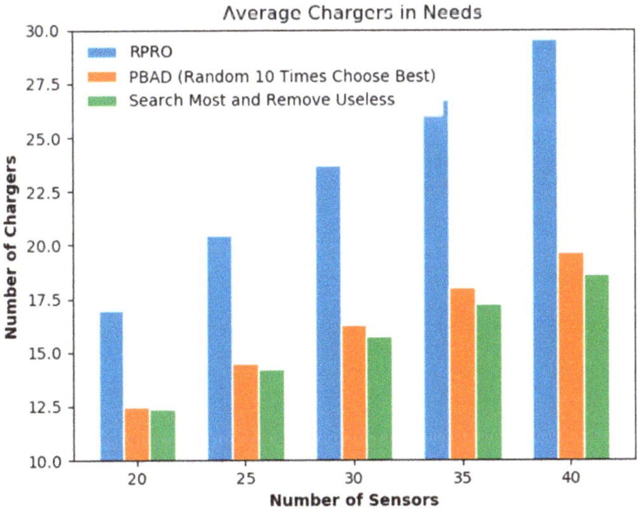

Figure 7. The average number of chargers.

4.3. Experimental Results of Execution Time

This experiment compares the relationship between the execution time of different algorithms and the number of sensors. Figure 8 shows the average execution time. As the RPRO algorithm selects the position and charging direction of the charger randomly and does not have too many calculations, it requires little execution time. However, the number of chargers required by the RPRO algorithm is very large. The PBAD algorithm randomly selects the charger position and repeats the operation multiple times to obtain the best results. Although it can obtain good results, the iterative operations are performed multiple times. As a result, the average execution time is more than that of SMRU and RPRO, and the execution time will increase as the number of recalculations increases. When SMRU selects the position that covers the largest sensor, it decides whether to compare multiple positions. Only when the number of sensors (candidate positions) around the position is more than three will it be executed. This is because when the number of candidate positions is less than three, regardless of the position selected, it does not affect the coverage results, and thus can reduce the execution time of many repeated calculations. Additionally, the SMRU algorithm only needs to be executed once to obtain the results, and it is not necessary to repeatedly take the best value like PBAD. Therefore, the execution time performance is much better than that of PBAD.

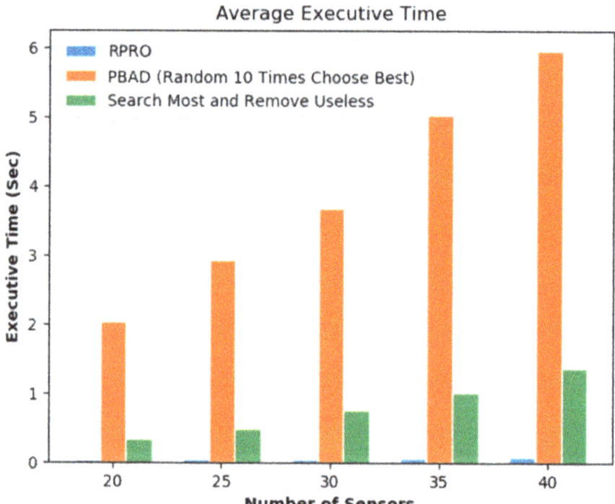

Figure 8. The average execution time.

4.4. Experimental Results for Different Charging Angles

This experiment compares the relationship between the size of the charging angle and the average number of chargers. The experimental angle ranges from 45° to 360°. The RPRO, PBAD, and SMRU algorithms were tested separately. The charging radius is 16 units, and the number of sensors is 40. The experiment found that the larger the charging angle, the fewer the number of chargers required. The different charger requirements for different algorithms are as follows: SMRU has the smallest, followed by PBAD, and the worst is RPRO. Additionally, for the same parameter settings of the same algorithm, there seems to be a lower demand limit on the number of charger requirements. As shown in Figure 9, it can be observed from the experimental results that there is no significant difference in the average number of chargers after 270° between PBAD and SMRU. However, the difference is significant when the angle is smaller than 270°.

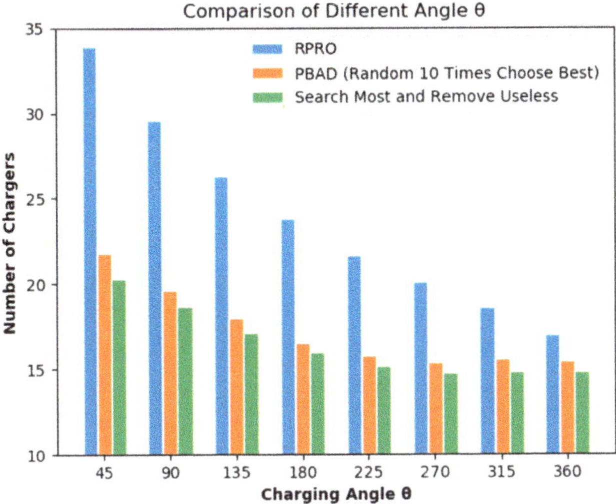

Figure 9. The comparison results of different charging angles.

4.5. The Comparison of the Execution Time of Charger Deployment and Modified Deployment Using SMRU

In the experiment, we measured the execution time it takes for the SMUR algorithm to adjust the position of chargers after sensor mobility. After the SMRU selects the charger position, this experiment randomly moves the sensors from 0 to 2 units to simulate the movement of the sensors. A comparison of the execution time of charger deployment and modified deployment using SMRU is shown in Figure 10.

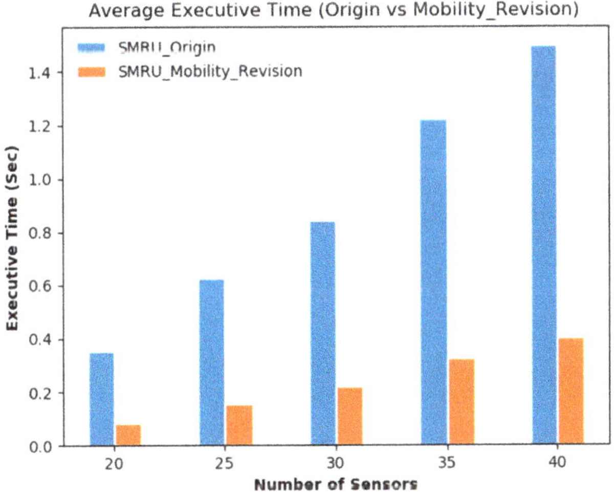

Figure 10. Comparison of the execution time of charger deployment and modified deployment using SMRU.

4.6. The Impact of Different Mobility Units

In this experiment, we first discuss the impact of different mobility units and average execution time for SMRU to adjust the position of chargers after sensor mobility. After the SMRU selects the charger position, this experiment randomly moves sensors from 2 to 10 units to simulate the movement of sensors. Due to SMRU does not recalculate the positions of all the chargers to find the global optimal solution after sensor mobility; it can reduce many execution times. The experimental results show that the execution time will increase with the number of sensors, and the increase in the mobility units of the sensor will increase the average execution time, as shown in Figure 11.

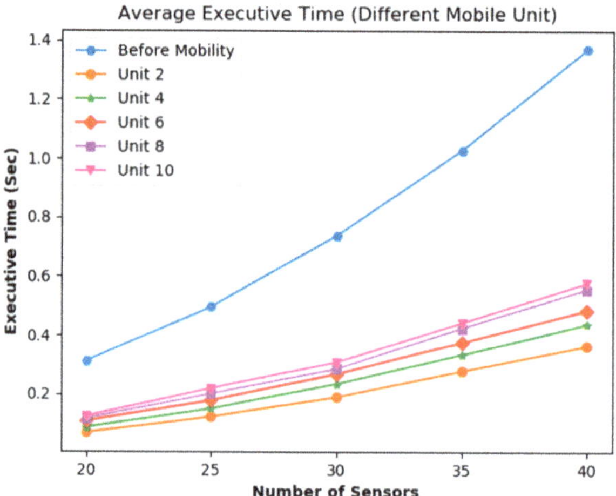

Figure 11. The impact of different mobility units and average execution time for search most and remove useless (SMRU) strategy to adjust the position of chargers after sensor mobility.

Next, we discuss the number of chargers used for SMRU to adjust the position of chargers after sensor mobility. Although SMRU does not recalculate the positions of all the chargers to find the global optimal solution after sensor mobility. However, the experimental results found that the number of chargers used does not increase too much after the sensor mobility. In addition, the number of chargers will increase with the number of sensors, and there is no obvious difference in the number of chargers used between different mobility units, as shown in Figure 12.

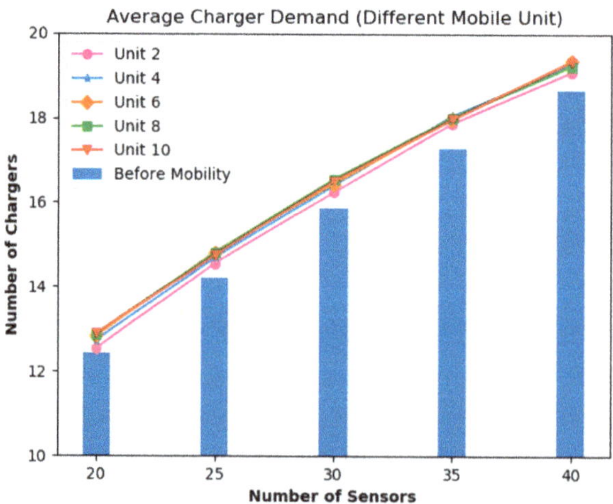

Figure 12. The impact of different mobility units and the number of chargers used for SMRU to adjust the position of chargers after sensor mobility.

4.7. Energy Consumption and Supplementary of Sink Nodes with Different Sink Rates

In this experiment, we discuss how to deploy enough chargers with different sink rates. In the experiment, an average of four sink nodes (SN) were deployed in a 100 × 100 unit experimental environment, with coordinates SN_1 (25, 25), SN_2 (25, 75), SN_3 (75, 25), SN_4 (75, 75). As all sensors need to send data to the sink node, it is assumed that the radius of the received data of the sink is 36 units in the experiment to cover the entire experimental environment. Every time a sensor transmits data to the sink node, it increases the sink energy consumption by 0.6 units. Each charger can charge 1 unit of energy to the sink. The total number of sensors deployed in the experimental environment was 20. This experiment is based on being able to replenish all the energy needed by the sink nodes, and deploying enough chargers around the sink node to replenish the power. Figure 13 shows the energy consumption and supplementary of sink nodes with different sink rates.

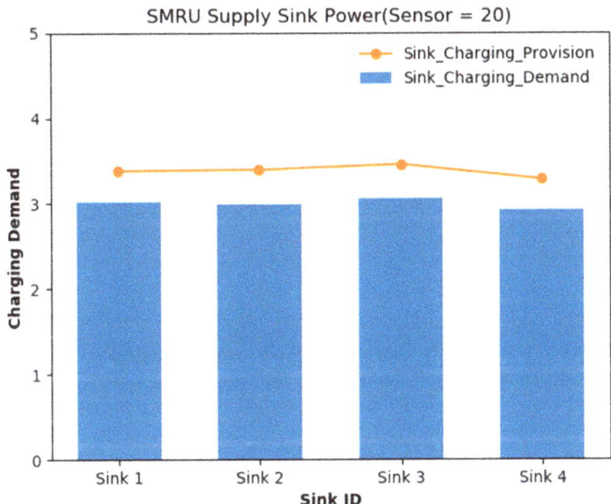

Figure 13. The energy consumption and supplementary of sink nodes with different sink rates.

4.8. The Impact of Different Densities on the Energy Requirements of Sink Nodes

This experiment compares the impact of different densities on the energy requirements of sink nodes, as shown in Figure 14. This experiment deploys four kinds of sink node density. The first case is 4C0S (C = Close, S = Separate) representing four sink nodes that have overlapping coverage. The second case is 2C2S representing two sink nodes that have overlapping coverage (Sink 1, Sink 2) and two isolated sink nodes (Sink 3, Sink 4). The third case is 3C1S represents three sink nodes that have overlapping coverage (Sink 1, 2, 3) and one isolated sink node (Sink 4). The fourth case is 0C4S, which means four sink nodes are isolated. The positions of sensor nodes are randomly deployed in the coverage of sink nodes. The experimental results found that the average energy requirements of sink nodes is almost the same in 0C4S and 4C0S situation, as shown in Figure 12. In the case of sink nodes that have overlapping coverage, sink nodes can share the processing of sensor nodes in the same coverage, and so the energy requirements of sink nodes will be shared and reduced. Therefore, the average energy requirements of sink nodes are the same as the four isolated sink nodes. In 2C2S and 3C1S situation, some sink nodes have overlapping coverage and some sink nodes are isolated. As isolated sink nodes have no other sink nodes in the coverage to help process the data from sensor nodes in the coverage, this causes an increase in the energy requirements of sink nodes, as shown in Figure 14.

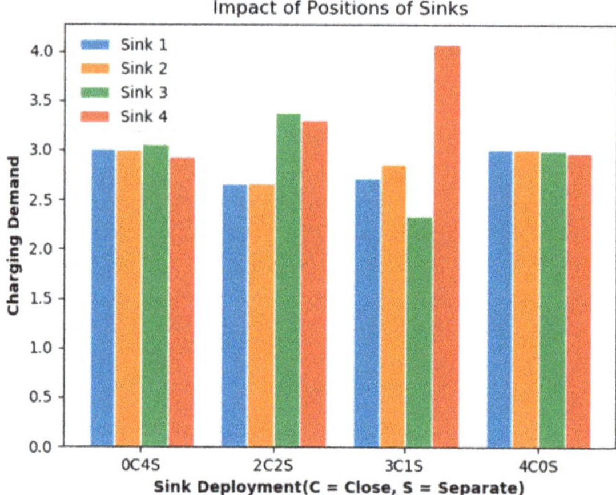

Figure 14. The impact of different densities on the energy requirements of sink nodes.

5. Conclusions

The deployment of WRSNs may require a large number of sensors. The use of fewer chargers in WRSNs to provide sufficient power for the sensors, reduce deployment costs, and maintain the operation of WRSNs is an important issue. Therefore, this study proposes an SMRU algorithm that uses fewer directional chargers to cover all sensors in WRSNs. In our proposed SMRU, the deployment position of the charger can be calculated in a short time, and all sensors in WRSNs are covered with a small number of chargers. Additionally, all sensor power is supplemented to avoid sensor interruption due to power consumption that causes WRSNs to fail and reduces the cost of deploying WRSNs. The performance of the proposed algorithm was compared with that of the RPRO and PBAD algorithms. Experimental results show that the average number of chargers used by the SMRU algorithm is fewer than that of RPRO and PBAD, and the execution time performance is much better than that of PBAD. In the future, we shall design a new algorithm that saves more chargers and computing time.

Author Contributions: Conceptualization, T.-L.L. and H.-Y.C.; methodology, T.-L.L. and H.-Y.C.; software, Y.-H.W.; validation, T.-L.L. and H.-Y.C.; formal analysis, T.-L.L.; investigation, T.-L.L.; resources, H.-Y.C.; writing—original draft preparation, T.-L.L. and H.-Y.C.; writing—review and editing, T.-L.L.; visualization, T.-L.L. and H.-Y.C.; funding acquisition, T.-L.L. and H.-Y.C. All authors have read and agreed to the published version of the manuscript.

Funding: This research was funded by Ministry of Science and Technology (MOST) of Taiwan, grant number 108-2321-B-415-003- and 109-2321-B-415-005-. MOST provided research funding and devices.

Conflicts of Interest: The authors declare no conflict of interest.

References

1. Pantazis, N.A.; Nikolidakis, S.A.; Vergados, D.D. Energy-Efficient Routing Protocols in Wireless Sensor Networks: A Survey. *IEEE Commun. Surv. Tutor.* **2013**, *15*, 551–591. [CrossRef]
2. Kausar, A.S.M.Z.; Reza, A.W.; Saleh, M.U.; Ramiah, H. Energy-Harvesting wireless sensor networks. *Renew. Sustain. Energy Rev.* **2014**, *38*, 973–989. [CrossRef]
3. Sample, A.P.; Meyer, D.T.; Smith, J.R. Analysis, Experimental Results, and Range Adaptation of Magnetically Coupled Resonators for Wireless Power Transfer. *IEEE Trans. Ind. Electron.* **2010**, *58*, 544–554. [CrossRef]
4. Global Wireless Charging Market 2016–2020. 2015. Available online: http://www.prnewswire.com/news-releases/global-wireless-charging-market-2016-2020-300196623.html (accessed on 31 July 2016).
5. Gilbert, D. WiFi Routers Can Wirelessly Charge Batteries, Cameras and Sensors. 2015. Available online: http://www.ibtimes.co.uk/wifi-routers-can-wirelessly-charge-batteries-cameras-sensors-1504987 (accessed on 31 July 2016).

6. Sendra, S.; Lloret, J.; Garcia, M.; Toledo, J.F. Power Saving and Energy Optimization Techniques for Wireless Sensor Neworks. *J. Commun.* **2011**, *6*, 439–459. [CrossRef]

7. Liao, J.-H. Wireless Charger Deployment Optimization for Wireless Rechargeable Sensor Networks. In Proceedings of the 2014 7th International Conference on Ubi-Media Computing and Workshops, Ulaanbaatar, Mongolia, 12–14 July 2013.

8. Horster, E.; Lienhart, R. Approximating optimal visual sensor placement. In Proceedings of the 2006 IEEE International Conference on Multimedia and Expo, Toronto, Canada, 9–12 July 2006.

9. Ai, J.; Abouzeid, A.A. Coverage by directional sensors in randomly deployed wireless sensor networks. *J. Comb. Optim.* **2006**, *11*, 21–41. [CrossRef]

10. Lin, T.L.; Li, S.L.; Chang, H.Y.J.E. A power balance aware wireless charger deployment method for complete coverage in wireless rechargeable sensor networks. *Energies* **2016**, *9*, 695. [CrossRef]

11. Dai, H.; Liu, Y.; Chen, G.; Wu, X.; He, T.; Liu, A.X.; Ma, H. Safe charging for wireless power transfer. *Tech. Rep.* **2017**, *25*, 3531–3544. [CrossRef]

12. He, S.; Chen, J.; Jiang, F.; Yau, D.K.Y.; Xing, G.; Sun, Y. Energy provisioning in wireless rechargeable sensor networks. *IEEE Trans. Mob. Comput.* **2013**, *2*, 1931–1942. [CrossRef]

13. Chiu, T.C.; Shih, Y.Y.; Pang, A.C.; Jeng, J.Y.; Hsiu, P.C. Mobility-Aware charger deployment for wireless rechargeable sensor networks. In Proceedings of the 2012 14th Asia-Pacific Network Operations and Management Symposium (APNOMS), Seoul, Korea, 25–27 September 2012.

14. Han, X.; Cao, X.; Lloyd, E.L.; Shen, C.C. Deploying directional sensor networks with guaranteed connectivity and coverage. In Proceedings of the 2008 5th Annual IEEE Communications Society Conference on Sensor, Mesh and Ad Hoc Communications and Networks, San Francisco, CA, USA, 16–20 June 2008.

15. Garey, M.R.; Johnson, D.S. *Computers and Intractibility: A Guide to the Theory of NP-Completeness*; W. H. Freeman & Co.: New York, NY, USA, 1979.

16. Mo, L.; Kritikakou, A.; He, S. Energy-Aware Multiple Mobile Chargers Coordination for Wireless Rechargeable Sensor Networks. *IEEE Internet Things J.* **2019**, *6*, 8202–8214. [CrossRef]

17. Tang, L.; Guo, H.; Wu, R.; Fan, B. Adaptive Dual-Mode Routing-Based Mobile Data Gathering Algorithm in Rechargeable Wireless Sensor Networks for Internet of Things. *Appl. Sci.* **2020**, *10*, 1821. [CrossRef]

18. Chen, J.; Yu, C.W.; Ouyang, W. Efficient Wireless Charging Pad Deployment in Wireless Rechargeable Sensor Networks. *IEEE Access* **2020**, *8*, 39056–39077. [CrossRef]

19. Sumi, F.; Dutta, L.; Sarker, F. Future with Wireless Power Transfer Technology. *J. Electr. Electron. Syst.* **2018**, *7*, 2332-0796.1000279.

Article

Enhancing the Performance of Energy Harvesting Sensor Networks for Environmental Monitoring Applications

Mahdi Zareei [1,*], Cesar Vargas-Rosales [1], Mohammad Hossein Anisi [2], Leila Musavian [2], Rafaela Villalpando-Hernandez [1], Shidrokh Goudarzi [3] and Ehab Mahmoud Mohamed [4,5]

[1]   Tecnologico de Monterrey, Escuela de Ingenieria y Ciencias, Monterrey 64849, Mexico
[2]   School of Computer Science and Electronic Engineering, University of Essex, Colchester CO4 3SQ, UK
[3]   Advanced Informatics School, Universiti Teknologi Malaysia Kuala Lumpur (UTM), Jalan Semarak, Kuala Lumpur 54100, Malaysia
[4]   Electrical Engineering Department, College of Engineering, Prince Sattam Bin Abdulaziz University, Wadi Addwasir 11991, Saudi Arabia
[5]   Electrical Engineering Department, Faculty of Engineering, Aswan University, Aswan 81542, Egypt
*   Correspondence: m.zareei@ieee.org

Received: 3 June 2019; Accepted: 11 July 2019; Published: 20 July 2019

Abstract: Fast development in hardware miniaturization and massive production of sensors make them cost efficient and vastly available to be used in various applications in our daily life more specially in environment monitoring applications. However, energy consumption is still one of the barriers slowing down the development of several applications. Slow development in battery technology, makes energy harvesting (EH) as a prime candidate to eliminate the sensor's energy barrier. EH sensors can be the solution to enabling future applications that would be extremely costly using conventional battery-powered sensors. In this paper, we analyze the performance improvement and evaluation of EH sensors in various situations. A network model is developed to allow us to examine different scenarios. We borrow a clustering concept, as a proven method to improve energy efficiency in conventional sensor network and brought it to EH sensor networks to study its effect on the performance of the network in different scenarios. Moreover, a dynamic and distributed transmission power management for sensors is proposed and evaluated in both networks, with and without clustering, to study the effect of power balancing on the network end-to-end performance. The simulation results indicate that, by using clustering and transmission power adjustment, the power consumption can be distributed in the network more efficiently, which result in improving the network performance in terms of a packet delivery ratio by 20%, 10% higher network lifetime by having more alive nodes and also achieving lower delay by reducing the hop-count.

Keywords: energy harvesting; wireless sensor network; environmental monitoring; clustering; energy efficiency; network lifetime; transmission power

1. Introduction

The increasing demand of smartness in our daily appliance and novel applications for more efficient use of the resources have led to a massive boom in the deployment of sensors. Due to technological advances, sensors are getting smaller and more powerful. However, the efficient managing of the small energy resource of the sensor is still a challenging task. Therefore, there have been several energy efficient algorithms and protocols proposed and put in experiments in many research papers and practical applications [1–4]. Apart from focusing on the efficient use of the limited power resource, harvesting energy from the ambient environment can be a promising solution for the aforementioned challenge. Energy harvesting cannot only extend the network lifetime in a sustainable manner, but it can also reduce the necessity of using batteries that pollute the environment due to leakage or when they are discarded [5].

While energy harvesting (EH) is not a new research topic, it has recently attracted more attention due to the massive implementation of sensors, for example in IoT applications. Imagine applications in which thousands of sensors are deployed to monitor the soil moisture in a massive farm for scheduling the watering, or in a forest for the early detection of massive fires. These massive implementations require sensors to work for a long time without a need of human intervention for maintenance. Thus, many researchers focus on overcoming research challenges in this area. Several methods of EH are developed that are powered by sunlight [6], vibration [7], heat [8], and electromagnetic waves [9–11], even from human body movement and heat [12]. However, the amount of the energy that can be harvested from these resources can be uncertain and unpredictable, which can cause difficulties in the design of different network stack protocols [13].

Similar to the effort to prolong the network lifetime in battery powered sensor networks, there are huge efforts on efficiently using the harvested power in energy harvesting sensor networks (EH-WSN), to have longer network lifetime and more reliable performance [14]. One of the ways that researchers have tried to achieve better energy efficiency in EH-WSN is to bring the clustering concept from WSN and try to remodel and adjust the idea to unique characteristics of EH-WSN. Basically, the clustering approach is a way to convert distributed decentralized networks into several small virtually centralized networks, and hence, bringing the benefits of a centralized architecture to a decentralized network [15]. In clustering, the network is divided into clusters, which consist of a cluster-head (CH), gateway nodes (GW) and member nodes (MN). CH is responsible for managing the MNs and processing the members' data, and communication between two adjacent clusters happens through GWs. In a cluster-based WSN, CHs, and GWs form the network backbone [16].

Clustering for EH-WSN can bring the similar benefits that it can bring to conventional WSN. However, considering the unique characteristic of EH-WSN such as an unpredictable amount of harvested power, clustering algorithms need to be carefully designed in a way to efficiently use the harvested power and at the same time try to deal with its uncertainty.

In this paper, we take a closer look at the effect of clustering for EH-WSN. The contributions of this paper are twofold: First is to study the effect of clustering on the performance of EH-WSN and second is to propose a way to improve the performance of the clustered network considering energy harvesting sensors. Therefore, we consider a decentralized network and study the effect of adjusting transmission power in the performance of the network in both flat (non-clustered) and clustered network. Given that the randomness in the transmission power can potentially affect the network topology and in some cases balance the power consumption among nodes with different harvesting ratios, the goal of this paper is to see how clustering can affect network performance in the existence of transmission power adjustment. For clustering, we used a simple energy-based clustering where each node connects to the neighbor with higher available energy. Thus, node A connects to node B as cluster head if node B has higher available power than node A. As cluster heads usually handle more traffic, by using this method, we can assure nodes with higher power become cluster head. However, any other clustering algorithm can be adjusted with the proposed transmission power control method. We show that, even by using a simple clustering concept, the network shows a fairer distribution of power consumption, which results in achieving better network end-to-end performance.

The rest of this paper is organized as follows: Section 2 presents the related work in clustering and power control for energy harvesting sensor networks. The system model is introduced in Section 3. The numerical results are presented and discussed in detail in Section 4. Section 5 presents the conclusion and future work.

2. Related Work

The main design challenge in battery-powered WSN is to prolong the network lifetime by using energy-efficient algorithms and protocols in medium access control (MAC) and Network layers [17,18]. Therefore, energy efficiency has been a very popular keyword in WSN literature. Researchers have been trying to tackle this issue by different means such as proposing methods that directly focus on energy conservation such as sleep scheduling [19], or by proposing energy efficient algorithms such as energy efficient routing, energy efficient medium access control protocol, using clustering [20], and even energy efficient coverage enhancement specially in mobile sensors case [21,22], which perform their classical task more energy efficiently. For example, as is discussed in Reference [19], by carefully selecting a certain number of nodes to be activated to cover a desired portion of a monitored area, network life-time can improve more than 80% in certain areas.

Clustering algorithms, as one of the important ways to reduce the network load and conserve energy, are widely studied in the literature [16]. Like conventional WSN, clustering approaches are widely applied in EH-WSN. However, in EH-WSN, as the amount of harvested energy is uncertain, the protocols should deal with unique challenges and specific characteristics. Several clustering algorithms are proposed for EH-WSN based on the famous Low Energy Adaptive Clustering Hierarchy (LEACH) protocol [23]. One of these extensions is Energy Potential LEACH (EP-LEACH), [24], which predicts the amount of energy that each node might harvest and tries to avoid using a low energy potential node as a cluster head. Using this method, EP-LEACH tries to construct more stable clusters and balance the power. Solar LEACH (sLEACH) is another extension of LEACH for energy harvesting sensors [25]. In sLEACH, the network consists of both solar and battery powered sensors. sLEACH gives higher priority to solar powered sensors with a high level of energy to become a cluster head.

Authors in Reference [26] proposed a centralized genetic-based unequal clustering algorithm for energy harvesting network (EHGUC). The main idea is that a base station creates clusters with different cluster sizes, in a way that the clusters that are closer to the base station will have smaller sizes, so as to make them consume less power. It uses several parameters such as the distance between nodes and energy harvesting rate as weighting parameters to select the cluster head. A gradient-based energy-efficient clustering (GEEC) packet forwarding mechanism is introduced in Reference [27]. Based on the relative position of the nodes and hop counts to the sink, a gradient model is constructed [27]. Cluster heads are selected in a distributed manner by considering the energy harvesting rate and the distance between a CH candidate and the center line of its circular ring. Thus, unequal clusters are formed to lower network overhead and balance network power consumption. The authors in Reference [28] proposed the use of energy harvesting sensors for relaying CH traffic. The concept of clustering is used in [29] to facilitate the routing in EH-WSN. Energy Neutral Clustering (ENC) [30], with the goal of providing perpetual network operation, groups the network into several clusters. It uses a mechanism called Cluster Head Group (CHG), which allows each cluster to have multiple cluster heads to distribute the traffic load. Using this method, the frequency of clustering and control message overhead is reduced. Moreover, using convex optimization techniques, the optimum number of clusters are computed. Another centralized clustering algorithm proposed is Reference [31] uses discrete particle swarm optimization (DPSO) for clustering where base station uses the status of all the nodes to modify the DPSO algorithm for clustering.

Apart from clustering and due to potentially unlimited harvested energy, several authors proposed various ideas about how to efficiently use the harvested power to achieve best performance [32]. One of the ways that is explored is by adjusting transmission power of the nodes. Optimal energy management policies for energy harvesting sensors are considered in Reference [33]. The discounted throughput is maximized over an infinite horizon, where queuing for data is also considered. In Reference [34], the authors proposed a solution to maintain the battery at a certain level in a network with time-varying battery recharging rate. In this work, two algorithms, namely, QuickFix and SnapIt, to compute the sampling rate and routes and to adapt to the rate, are respectively proposed. In Reference [35], to maximize the successful transmission and control energy–error probability tradeoff, Markov

Decision Processes (MDP) have been used to choose between multiple transmissions. The authors in Reference [36] developed a discrete time Markov model for different energy profile and storage capabilities to analyze node outage probabilities. A complete information Markov decision process model is developed in Reference [37] to characterize sensor's battery recharge/discharge process to achieve optimal transmit policies. In Reference [38], the authors used an infinite time–horizon Markov decision process (MDP) to formulate the power control. They obtained a closed-form threshold-based optimal power control solution for EH-WSN. The authors in Reference [39] formulate the power control as a stochastic optimization problem and solve the optimization problem using Lyapunov techniques. Without requiring historical knowledge of energy harvesting only based on the channel fade conditions and current energy state of the battery, their proposed algorithm adjusts the transmission power.

To the best of our knowledge, there is a limited number of contributions in the literature about clustering and power control on energy harvesting sensor networks. However, there are some similar works on conventional battery powered sensors. Li et al. [40] proposes an Energy-Efficient Unequal Clustering (EEUC) mechanism, which partitions nodes into unequal cluster sizes in a way that clusters closer to the base station have smaller sizes than those clusters that are farther away from the base station. Using this method, there would be less power consumption due to intra-cluster data forwarding for cluster heads closer to the base station.

Rehman Khan et al. [41] proposed two routing protocols, namely power-controlled routing (PCR) and enhanced power-controlled routing (EPCR), where the basic idea of both is nodes only change their transmission power when they want to transfer their data to the cluster-heads. Thus, nodes that are closer to the cluster-head can transmit with lower power, which eventually results in less power consumption. The authors in Reference [42] propose a clustering algorithm and a transmission power control mechanism for an ad hoc network, aiming to reduce the impacts of mobility and adaptive transmission power. The idea is that the cluster-head adjusts its transmission power to the changes in its cluster. For example, if CH receives a weak signal from a sensor or a report about unusual error rate, it increases its transmission power to the maximum to cover all the nodes in its vicinity. In another work, the authors in Reference [43] propose a power aware multihop routing protocol for a cluster based sensor network, where the objective is to minimize the power consumption of CH, then it sends the data in a power-aware multihop manner to the base station through a quasi-fixed route (QFR), where nodes vary their transmission power based on the distance to the receiver.

This paper does not aim to propose a novel clustering scheme for energy harvesting sensors, nor does it intend to propose a novel power control scheme for WSN. As energy harvesting is spatio-temporal dependent, nodes might harvest energy with different ratios. Considering this uncertainty in available energy, the main goal of this paper is to study how transmission power control can be effective in a clustered or structured network in EH-WSN.

In our earlier work, we proposed a transmission power control for an unstructured/flat network, where each node sends their data to the base-station that is located in the center of the network using multihop forwarding. We have concluded that the power control can help the load balancing among the sensors in directional communication from sensors to base-station, which results in the better end-to-end performance. However, in this paper, we take a step further to study how power control can affect the performance of the network where there is more dynamicity in the transmission among the nodes. Therefore, we considered a network without any central entity, where sensors communicate together in multihop fashion bi-directionally. Thus, at one round, a node can be a forwarder and at another, it can be the receiver of the packet. We study this dynamicity and randomness in an unstructured network, and then we compare it with a structured network.

The main contributions of this paper can be summarized as follows:

- Studying the effect of power control in an uninstructed network where nodes communicating together in a random multi-hop fashion.
- Studying the adaptive transmission power control effect in the clustered network.
- Comparing the proposed method with some existing work in terms of network lifetime.

3. System Model

We assume there are N sensors deployed non-uniformly in the network, where each sensor is equipped with an energy harvesting unit that is capable of harvesting energy from its ambient environment, e.g., solar, radio wave. The harvested energy is stored in a storage device such as a supercapacitor to supply power for data transmission.

The system operates in a discrete slotted time $t \in \{0, 1, 2, \dots \}$ and all operations are performed in each slot with duration Δt. Let $E_h(t)$ denote the amount of harvested energy, and $E_c(t)$ denote the amount of consumed energy both during time slot t. Let $\rho_c \in (0, 1]$ and $\rho_d \in [1, \infty)$ denote charging and discharging efficiency, respectively. Thus, the remaining energy at the beginning of slot $t + 1$ is denoted by $E_r(t+1)$ and is defined as

$$E_r(t + 1) = \rho_d E_r(t) + (\rho_c E_h(t) - E_c(t)) \forall t \tag{1}$$

In this work, for simplicity, the charging and discharging rate of the supercapacitor are considered negligible, thus $\rho_c \approx \rho_d \approx 1$. Considering E_{max} as the maximum storage capacity, thus $0 \leq E_r(t) \leq E_{max}$.

In a real-world application, the available harvested energy would be spatio-temporal dependent, which means it varies based on the location of the sensors and time. To emulate this effect, we divide the network in different regions R_i where nodes located in each region harvest with the same probability η_i that is a uniformly distributed random number. Let $E_{h,max}$ and $E_{h,min}$ denote the maximum and minimum harvesting rate, which should satisfy $E_{h,min} \leq \eta_i \leq E_{h,max}$. In this paper $[E_{h,min}, E_{h,max}] = [0.5,1]$ is considered. In a solar powered sensor, η_i value 0.5 is considered to mimic a partially clouded area where nodes in that region can harvest with half the value of $E_{h,max}$, and $\eta_i = 1$ represents a clear sky where nodes can harvest energy with the highest ratio.

Based on the remaining energy of each node $i \in N$, $E_{r,i}$, a node i can be in three different states, namely stable, medium, and critical, denoted as $N_{s_i,s}$, $N_{s_i,m}$ and $N_{s_i,c}$, respectively. Thus, for node i, it becomes

$$N_{s_i} = \begin{cases} N_{s_i,s} & \text{defines as} \quad E_{r_i} \geq \alpha \\ N_{s_i,m} & \text{defines as} \quad \beta \leq E_{r_i} < \alpha \\ N_{s_i,c} & \text{defines as} \quad \theta \leq E_{r_i} < \beta \end{cases}, \tag{2}$$

where α, β, and θ are power parameters that are tunable based on network requirements.

When using transmission power control, for each node, three different transmission power levels are considered, namely, normal, medium, and high, denoted by $N_{tp,n}$, $N_{tp,m}$, and $N_{tp,h}$, respectively. A node switches between these three modes based on its own power level and their neighboring nodes' power condition, as depicted in Equation (3). Therefore, each node adapts its transmission power based on the other nodes' condition in its surrounding environment. Thus, for node i, we have

$$N_{tp_i} = \begin{cases} N_{tp_i,n} & \text{if} \quad N_{s_i,c} \;\vee\; N_{s_j,s}, \forall j \in N_{ngh,i} \\ N_{tp_i,m} & \text{if} \quad (N_{s_i,s} \vee N_{s_i,m}) \wedge \; N_{s_j,c}, \forall j \in N_{ngh,i} \\ N_{tp_i,h} & \text{if} \quad N_{s_j,s} \wedge \; N_{s_j,c}, \forall j \in N_{ngh,i} \end{cases}, \tag{3}$$

where $N_{ngh,i}$ is the set of node i neighbors in which $N_{ngh,i} \in N$.

In each round (t), a certain number of the nodes N_d generates and transmits their event to a random destination (another node inside the network) over a time-varying wireless channel. The packets are forwarded using neighboring nodes towards the destination in a greedy manner using the shortest hop distance. Therefore, each node might work either as a relay or as a data source or data destination. Therefore, at each round t, node energy consumption is calculated as the maximum power required to sense its own packet or summation of all relayed packets at each round, which can be expressed as

$$F_{c,i}(t) = \max\left\{ E_{t_i}(t), \sum_{k \in N_{ngh,i}} P_{f,k}(t) \right\}, \tag{4}$$

where E_{t_i} is the energy node i consumes for transmitting its sensing packet and $P_{f,k}$ is the of number of forwarded packets from node i neighbors.

Each node can communicate with the neighboring nodes in its coverage region. A circular region centered on a node is referred to as coverage region of a node, where the radius of the circle can vary, and it is determined by each node transmission power and pathloss model. There are several pathloss models that are used in the literature. In this paper we model the pathloss using log-distance model [44] as follows

$$\overline{P_{R_{dB}}}(d) = P_{T_{dB}} - 10n \log_{10}(d). \tag{5}$$

where $\overline{P_{R_{dB}}}(d)$ is the average received power in dB at separation distance d, $P_{T_{dB}}$ is the transmission power in dB, and n is the path loss exponent. Considering the normal distribution, Equation (5) can be written as

$$P_{R_{dB}}(d) = P_{T_{dB}} - 10n \log_{10}(d) + X_\sigma, \tag{6}$$

where $X_\sigma \sim \mathcal{N}(0, \sigma)$ is a Gaussian random variable with zero mean and standard deviation σ in dB. Thus, $P_{R_{dB}}(d)$ is also Gaussian distributed, and can be formulated as

$$P_{R_{dB}}(d) \sim \mathcal{N}\left(P_{T_{dB}} - 10n \log_{10}(d), \sigma\right). \tag{7}$$

When the receiver sensitivity is lower than the received power, the received signal can be readable at the receiver. Therefore, as $P_{R_{dB}}(d)$ is Gaussian distributed, we use the Q-function to calculate the probability of $P\left(P_{R_{dB}}(d) > \gamma_{dB}\right)$ (successful received signal) by calculating the tail probability of the standard normal distribution [44]. Thus, we have

$$P\left(P_{R_{dB}}(d) > \gamma_{dB}\right) = Q\left(\frac{\gamma_{dB} - \left(P_{T_{dB}} - 10n \log_{10}(d)\right)}{\sigma}\right). \tag{8}$$

Thus, each packet will be successfully delivered to the next hop with a probability of $P\left(P_{R_{dB}}(d) > \gamma_{dB}\right)$ at each round. Packets might not reach the destination due to disconnected links, shortage of remaining energy in nodes, or channel conditions. In those cases, the packet will be dropped and measurement outage occurs.

4. Performance Evaluation

As it is costly to have testbed or implementation in a large scale, we have decided to study the performance of the network using MATLAB, where various conditions and different sets of simulations have been conducted using real life parameter values and features that would help us understand the behavior in the design and planning phase of the network. 100 CR sensor nodes deployed non-uniformly in a 1000 m × 1000 m area, which can emulate a 100 hectares' forest. We assume each node is equipped with a supercapacitor such as 3FNESSCAP [45], as energy storage for harvested energy. At the initial stage, the supercapacitor is considered fully charged and it can hold up enough power for 300 events. Simulation runs for 1000 rounds and in each round 30 randomly chosen sources transfer a packet to 30 randomly chosen destinations using the shortest available path, where several performance metrics such as average packet delivery ratio (PDR), average hop-count, and consumed power are studied. To achieve reliable results, reducing the effect of randomness and obtaining sufficient confidence, we run the simulation with 50 different network setups. Results are averaged with a 90% confidence interval. Simulation parameters are summarized in Table 1.

Table 1. Simulation parameters.

Parameter	Value
Number of nodes	100, 200
Network size	1000 m × 1000 m
Transmission range	Default = 100, Δ = 150, δ = 200 m
Power parameters	A = 50%, β = 25%, θ = 10%
Capacitor	300 energy units
Packet rate	30 packets per round
Energy harvesting rate	6 units per round
Consumption rate	1,3,10 units per round depending on transmission power
Noise Floor	0, 10, 20, 30, 40 dB
Simulation run	50 rounds
Confidence interval	90%

4.1. Non-Clustered Network Performance Evaluation

In the first simulation, a flat (non-clustered) network is considered where each node can communicate with the neighboring nodes within its transmission range. The result of this simulation is depicted in Figures 1–4. Figure 1 depicts an example of a deployed network, with the heatmap highlighting the amount of available energy in the deployment field, where x and y axis defining the size of the field. The figure shows the case with power control ON and OFF. Each case is shown with four subgraphs that are obtained as time evolves in the simulation. As can be observed, the nodes in the center of the network that usually relay more traffic deplete their energy faster than those nodes in the outskirts of the network. However, using adaptive power control, it can be observed that energy is distributed more uniformly than the case without power control.

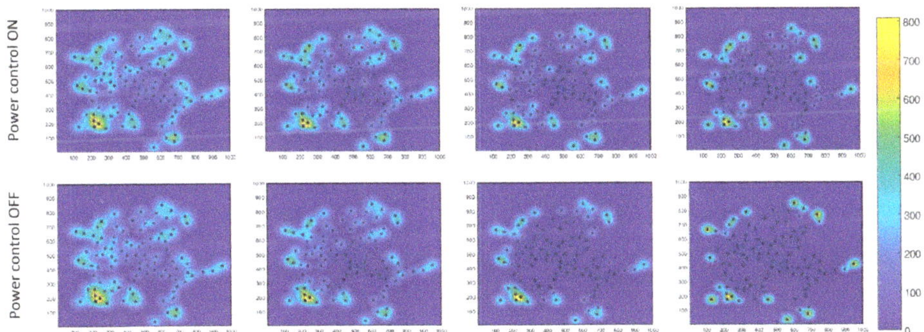

Figure 1. Network power level heatmap demonstrating the changes in the nodes power level in four stages of the running network.

Figure 2 depicts the effect of power control on the percentage of inactive and disconnected nodes. Inactive nodes are defined as those nodes that ran out of power and are deactivated until they harvest enough energy to go back to sensing state. Meanwhile, disconnected nodes are defined as those nodes that have enough power to operate, however, they got disconnected from the rest of the network because of the broken links due to the inactive nodes. As can be observed from Figure 2a, when power control is ON, the network shows a slightly lower number of inactive nodes. The reason is that, by using power control, the nodes balance the energy consumption in a way that they try to reduce the traffic load on low power nodes indirectly forcing them to consume less energy. By increasing the number of nodes to 200, we can see more nodes that are running out of energy, where in round 1000, it reaches 9% and 13%, with and without power control, respectively. The reason behind this sharp increase in the number of inactive nodes can be due to longer hop routes from each source to destination, which causes more energy consumption. The effect of those inactive nodes on the network can be seen in Figure 2b, where the inactive nodes cause a part of the network to be disconnected from the rest. As it is shown earlier in Figure 1, nodes that are usually handling more traffic deplete their energy faster and run out of batteries. As those nodes are usually located in the central area and work as bridges between two sections of the network, their deaths cause network partitioning and disconnected nodes. However, as is shown, using adaptive power control, the network stays connected even in the presence of the same number of inactive nodes as nodes find new connections in the network since nodes can increase their transmission power. By increasing the number of nodes to 200, the network shows lower disconnected nodes due to more connections available that each node has. However, the pattern is still the same.

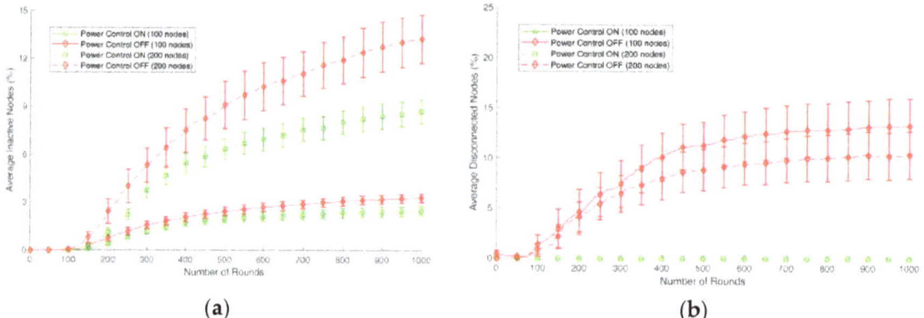

Figure 2. Network performance in terms of (**a**) number of inactive nodes and (**b**) number of disconnected nodes.

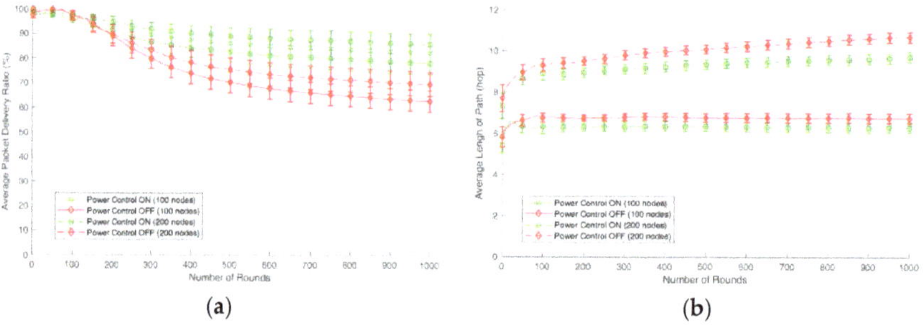

Figure 3. Network performance in terms of (**a**) average packet delivery ratio and (**b**) average length of path.

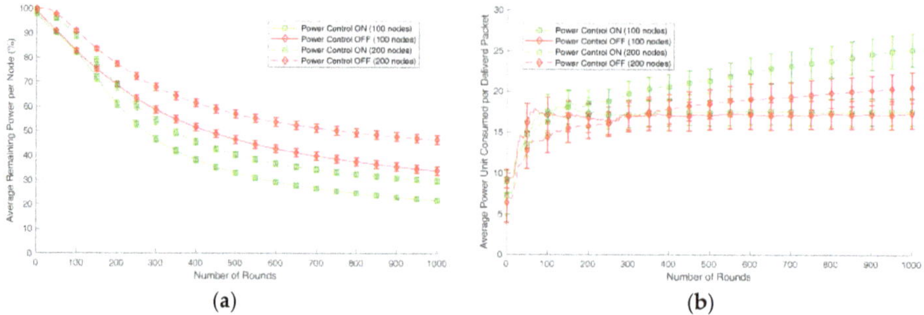

Figure 4. Network performance in terms of (**a**) average remaining power per node and (**b**) average power consumed per delivered packet.

The effect of power control on network end-to-end performance is depicted in Figure 3. The figure reveals that using power control, the network achieves around 20% higher PDR ratio compared to without the use of it, in both 100 and 200 nodes scenarios. The main reason is the number of disconnected nodes from the network in which they cannot reach the rest of the network as observed in Figure 2b. In terms of average hop count depicted in Figure 3b, due to higher transmission range when power control is ON, it shows slightly lower average hop count, which can result in a lower delay. By increasing the number of nodes to 200, the network expands and shows, on average, 2 hop longer routes, but the same behavior.

The network performance in terms of power consumption is shown in Figure 4. It can be observed from Figure 4a that using adaptive power control, the network consumes more energy (as it is expected) and the average remaining energy per node is less than that when not using power control. This is due to the increase in transmission power of some nodes to keep the network connected. It can be seen from Figure 4b that using adjustable transmission power, the network performs the same with and without power control when having 100 nodes, but when having 200 nodes, the network consumes more energy for each packet while using the power control. The reason is again the extra power used in the power control case to keep the network connected. However, considering that it achieves 20% extra PDR (see Figure 3a), it is a reasonable tradeoff.

4.2. Clustered Network Performance Evaluation

Figures 5–8 show the performance of the network where nodes are divided into clusters based on the conditions explained in the system model. Similar to Figure 1, Figure 5 shows the changes in power intensity of a clustered network in different stages. As is shown, using adaptive power control, the network balances the energy consumption in different parts of the network. Compared to Figure 1, it can be observed that, with clustering, the network achieves better energy distribution that results in better performance.

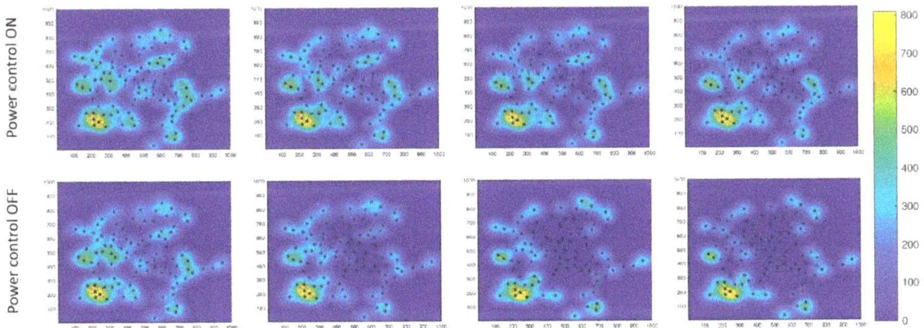

Figure 5. Clustered network power level heatmap demonstrating the changes in the nodes power level in four stages of the running network.

Figure 6 shows the performance of the network in terms of the lifetime of nodes. As can be seen from Figure 6a from round 200, some nodes start running out of batteries. As can be seen, using an adaptable power control network shows slightly lower number of inactive nodes. Like the flat network, in cluster network increasing the number of nodes to 200, we observe more inactive nodes. However, using clustering network shows around 3% less inactive nodes compared to a flat network. Similar to Figure 2b, in Figure 6b, it can also be observed that, due to those inactive nodes, several other nodes get disconnected from the network. However, by adjusting the power control to the network condition, nodes stay connected. Comparing Figures 2 and 6, we can see how clustering in the network improves the performance and we observe a lower number of inactive and disconnected nodes.

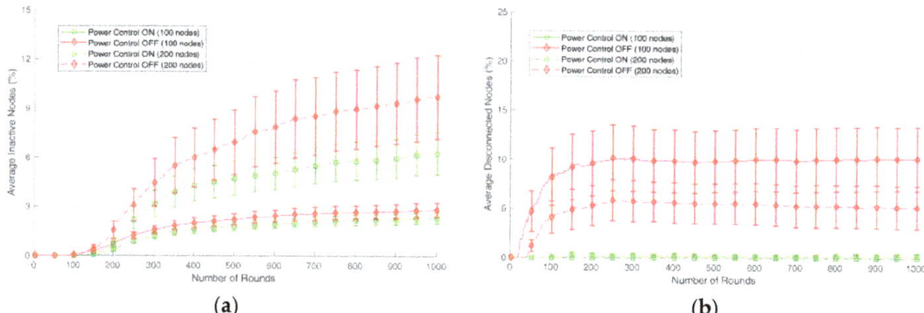

Figure 6. Network performance in a clustered network in terms of (**a**) number of inactive nodes and (**b**) number of disconnected nodes.

Figure 7 depicts the network end-to-end performance. As can be seen in Figure 7a, without using power control network achieves around 70% with 100 nodes and 80% with 200 nodes, which shows improvement compared to the network without clustering that is around 62% and 70% in the 100 and 200 nodes case, respectively (Figure 3a). The reason behind this improvement is the lower disconnected nodes in the network. While using adaptable power control, the network reaches higher PDR and the performance is similar with the network without clustering and it shows around 80% and more than 90% PDR in the 100 and 200 nodes case, respectively. In terms of average hop-count depicted in Figure 7b, the network shows almost the same pattern with a slight increase in hop count compared to a network without clustering.

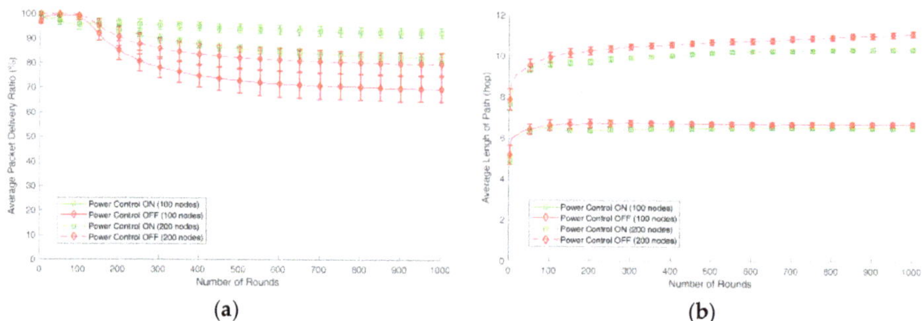

Figure 7. Network performance in a clustered network in terms of (**a**) average packet delivery ratio and (**b**) average length of path.

Figure 8 shows the network performance in terms of power consumption. It can be seen that, due to an increase in transmission power using adaptive power control network, it consumes more power on average and it drops to nearly 35% and 45% for the 100 and 200 nodes, respectively. However, comparing Figure 8a with Figure 4a, it can be observed that using clustering network results in more than a 10% increase in remaining power per node than when not using it. In terms of packet cost, with and without power control, the network shows the same performance and it consumes 15 power units on average, where this value is 17 units without clustering for 100 nodes. Using 200 nodes, the network shows the same pattern with an increase in packet cost that reaches 20 units per packet.

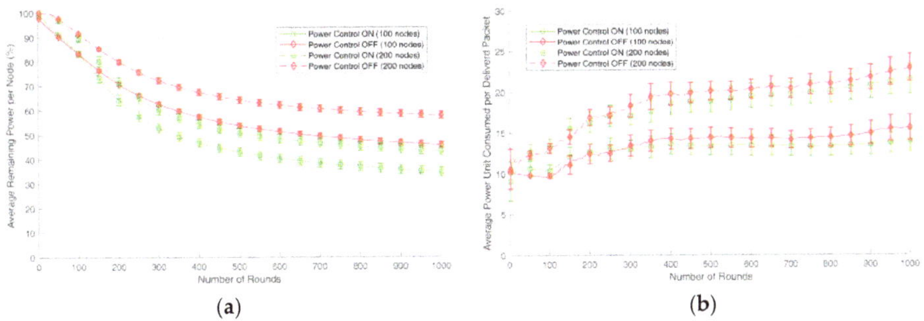

(a)   (b)

Figure 8. Network performance in a clustered network in terms of (**a**) average remaining power per node and (**b**) average power consumed per delivered packet.

A comparison of the proposed protocol with three well-known protocols energy-efficient unequal clustering (EEUC), hybrid, energy-efficient, distributed clustering (HEED) [46], and low energy adaptive clustering hierarchy (LEACH) [47] has been performed in order to study the performance of the proposed protocol. The result of the comparison is depicted in Figure 9. As can be seen from the figure, the proposed protocol start losing some nodes around round 300, while other protocols start performing better. The reason behind is the sudden increase in the transmission power of some nodes in the proposed protocol. However, the rate of dying nodes starts to slow down by round 500. Other protocols maintain most of the nodes' power, but they face a sudden decrease in the number of alive nodes, as by then, most of the nodes lost their high percentage of power. As can be seen, by round 700 to 800, most of the nodes are dead in the other protocol, while the proposed protocol shows better performance having around 50% of the nodes alive. The reason is that nodes are relaying lower traffic due to the loss of some connections.

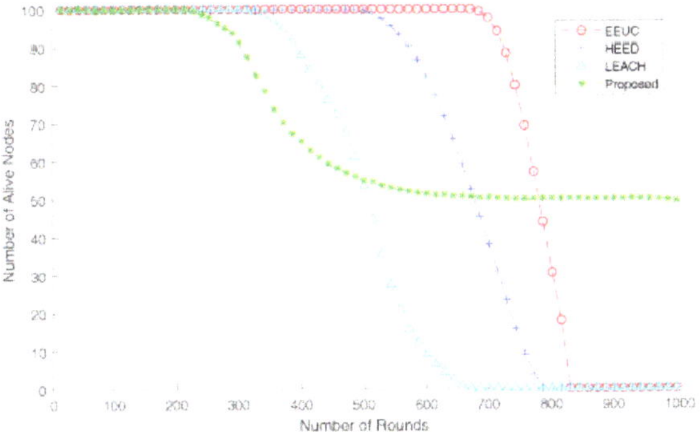

Figure 9. Performance comparison regarding the network lifetime.

5. Conclusions

Energy harvesting is a sustainable way to prolong lifetime of a sensor network, which enable us to target various applications that are not feasible using battery-powered sensors. However, the uncertainty in the amount of available energy poses design challenges. This paper studied the effect of using adaptive transmission power control for energy harvesting sensors in both clustered and non-clustered networks. The adaptive transmission power control adjusts the transmission power of each individual node independently based on the node residual power and its neighboring nodes' energy condition. Once a node energy level goes below a predefined threshold level, the neighboring nodes increase their transmission power control based on their residual power accordingly, to reduce the traffic load on the low energy node. Using this method, power can be consumed more uniformly among the nodes in the network. The numerical results indicate that the proposed adaptive transmission power control improves the performance of non-clustered networks. We also have proven that a basic power control can still improve the clustered network performance further and help the network to achieve better end-to-end performance. In this paper, we have used a simple energy-based clustering, greedy routing, and relatively simple power control mechanism to prove the efficiency. However, in the future other clustering and routing algorithms are going to be considered, which we believe results in further improvement of network performance metrics.

Author Contributions: Conceptualization, M.Z. and C.V.-R.; methodology, M.Z., C.V.-R.; validation M.Z., C.V.-R., M.H.A.; writing—original draft preparation, M.Z.; writing—review and editing, C.V.-R., M.H.A., L.M., E.M.M., R.V.-H., S.G.; visualization, M.Z.; supervision, C.V.-R.; funding acquisition, C.V.-R., M.H.A.

Funding: This research was funded by the Royal Society of the UK under grant IES\R3\170342 and the SEP-CONACyT Research Project under grant 255387, the School of Engineering and Sciences and the Telecommunications Research Group at Tecnologico de Monterrey.

Conflicts of Interest: The authors declare no conflict of interest.

References

1. Antolín, D.; Medrano, N.; Calvo, B.; Pérez, F. A wearable wireless sensor network for indoor smart environment monitoring in safety applications. *Sensors* **2017**, *17*, 365. [CrossRef] [PubMed]
2. Kurt, S.; Yildiz, H.U.; Yigit, M.; Tavli, B.; Gungor, V.C. Packet size optimization in wireless sensor networks for smart grid applications. *IEEE Trans. Ind. Electron.* **2017**, *64*, 2392–2401. [CrossRef]
3. Park, P.; Di Marco, P.; Johansson, K.H. Cross-layer optimization for industrial control applications using wireless sensor and actuator mesh networks. *IEEE Trans. Ind. Electron.* **2017**, *64*, 3250–3259. [CrossRef]
4. Spirjakin, D.; Baranov, A.; Akbari, S. Wearable Wireless Sensor System with RF Remote Activation for Gas Monitoring Applications. *IEEE Sens. J.* **2018**, *18*, 2976–2982. [CrossRef]
5. Dehghani-Sanij, A.; Tharumalingam, E.; Dusseault, M.; Fraser, R. Study of energy storage systems and environmental challenges of batteries. *Renew. Sustain. Energy Rev.* **2019**, *104*, 192–208. [CrossRef]
6. Mekikis, P.V.; Kartsakli, E.; Antonopoulos, A.; Alonso Zárate, L.G.; Verikoukis, C. Connectivity analysis in clustered wireless sensor networks powered by solar energy. *IEEE Trans. Wirel. Commun.* **2018**, *17*, 2389–2401. [CrossRef]
7. Wei, C.; Jing, X. A comprehensive review on vibration energy harvesting: Modelling and realization. *Renew. Sustain. Energy Rev.* **2017**, *74*, 1–18. [CrossRef]
8. Allmen, L.; Bailleul, G.; Becker, T.; Decotignie, J.-D.; Kiziroglou, M.; Leroux, C.; Mitcheson, P.; Müller, J.; Piguet, D.; Toh, T. Aircraft strain WSN powered by heat storage harvesting. *IEEE Trans. Ind. Electron.* **2017**, *64*, 7284–7292. [CrossRef]
9. Wang, C.; Li, J.; Yang, Y.; Ye, F. Combining solar energy harvesting with wireless charging for hybrid wireless sensor networks. *IEEE Trans. Mob. Comput.* **2018**, *17*, 560–576. [CrossRef]
10. Lin, C.; Wu, Y.; Liu, Z.; Obaidat, M.S.; Yu, C.W.; Wu, G. GTCharge: A game theoretical collaborative charging scheme for wireless rechargeable sensor networks. *J. Syst. Softw.* **2016**, *121*, 88–104. [CrossRef]
11. Lin, C.; Wu, G.; Obaidat, M.S.; Yu, C.W. Clustering and splitting charging algorithms for large scaled wireless rechargeable sensor networks. *J. Syst. Softw.* **2016**, *113*, 381–394. [CrossRef]
12. Zhou, M.; Al-Furjan, M.S.H.; Zou, J.; Liu, W. A review on heat and mechanical energy harvesting from human–Principles, prototypes and perspectives. *Renew. Sustain. Energy Rev.* **2018**, *82*, 3582–3609. [CrossRef]
13. Anisi, M.H.; Abdul-Salaam, G.; Idris, M.Y.I.; Wahab, A.W.A.; Ahmedy, I. Energy harvesting and battery power based routing in wireless sensor networks. *Wirel. Netw.* **2017**, *23*, 249–266. [CrossRef]
14. Zareei, M.; Vargas-Rosales, C.; Villalpando-Hernandez, R.; Azpilicueta, L.; Anisi, M.H.; Rehmani, M.H. The effects of an Adaptive and Distributed Transmission Power Control on the performance of energy harvesting sensor networks. *Comput. Netw.* **2018**, *137*, 69–82. [CrossRef]
15. Zareei, M.; Islam, A.M.; Vargas-Rosales, C.; Mansoor, N.; Goudarzi, S.; Rehmani, M.H. Mobility-aware medium access control protocols for wireless sensor networks: A survey. *J. Netw. Comput. Appl.* **2018**, *104*, 21–37. [CrossRef]

16. Zeb, A.; Islam, A.M.; Zareei, M.; Al Mamoon, I.; Mansoor, N.; Baharun, S.; Katayama, Y.; Komaki, S. Clustering analysis in wireless sensor networks: The ambit of performance metrics and schemes taxonomy. *Int. J. Distrib. Sens. Netw.* **2016**, *12*, 4979142. [CrossRef]

17. Zareei, M.; Islam, A.M.; Mansoor, N.; Baharun, S.; Mohamed, E.M.; Sampei, S. CMCS: A cross-layer mobility-aware MAC protocol for cognitive radio sensor networks. *EURASIP J. Wirel. Commun. Netw.* **2016**, *2016*, 160. [CrossRef]

18. Zareei, M.; Taghizadeh, A.; Budiarto, R.; Wan, T.-C. EMS-MAC: Energy efficient contention-based medium access control protocol for mobile sensor networks. *Comput. J.* **2011**, *54*, 1963–1972. [CrossRef]

19. Mostafaei, H.; Montieri, A.; Persico, V.; Pescapé, A. A sleep scheduling approach based on learning automata for WSN partialcoverage. *J. Netw. Comput. Appl.* **2017**, *80*, 67–78. [CrossRef]

20. Uddin, J.; Ghazali, R.; Deris, M.M. An empirical analysis of rough set categorical clustering techniques. *PLoS ONE* **2017**, *12*, e0164803. [CrossRef]

21. Mohamed, S.M.; Hamza, H.S.; Saroit, I.A. Coverage in mobile wireless sensor networks (M-WSN): A survey. *Comput. Commun.* **2017**, *110*, 133–150. [CrossRef]

22. Khan, A.; Abdullah, A.; Anisi, M.; Bangash, J. A comprehensive study of data collection schemes using mobile sinks in wireless sensor networks. *Sensors* **2014**, *14*, 2510–2548. [CrossRef]

23. Heinzelman, W.B.; Chandrakasan, A.P.; Balakrishnan, H. An application-specific protocol architecture for wireless microsensor networks. *IEEE Trans. Wirel. Commun.* **2002**, *1*, 660–670. [CrossRef]

24. Xiao, M.; Zhang, X.; Dong, Y. An effective routing protocol for energy harvesting wireless sensor networks. In Proceedings of the 2013 IEEE Wireless Communications and Networking Conference (WCNC), Shanghai, China, 7–10 April 2013; pp. 2080–2084.

25. Voigt, T.; Dunkels, A.; Alonso, J.; Ritter, H.; Schiller, J. Solar-aware clustering in wireless sensor networks. In Proceedings of the ISCC 2004, Ninth International Symposium on Computers and Communications (IEEE Cat. No.04TH8769), Alexandria, Egypt, 28 June–1 July 2004; Volume 1, pp. 238–243.

26. Wu, Y.; Liu, W. Routing protocol based on genetic algorithm for energy harvesting-wireless sensor networks. *IET Wirel. Sens. Syst.* **2013**, *3*, 112–118. [CrossRef]

27. Wu, D.; He, J.; Wang, H.; Wang, C.; Wang, R. A hierarchical packet forwarding mechanism for energy harvesting wireless sensor networks. *IEEE Commun. Mag.* **2015**, *53*, 92–98. [CrossRef]

28. Zhang, P.; Xiao, G.; Tan, H.-P. Clustering algorithms for maximizing the lifetime of wireless sensor networks with energy-harvesting sensors. *Comput. Netw.* **2013**, *57*, 2689–2704. [CrossRef]

29. Alrajeh, N.A.; Khan, S.; Lloret, J.; Loo, J. Secure routing protocol using cross-layer design and energy harvesting in wireless sensor networks. *Int. J. Distrib. Sens. Netw.* **2013**, *9*, 374796. [CrossRef]

30. Peng, S.; Wang, T.; Low, C. Energy neutral clustering for energy harvesting wireless sensors networks. *Ad Hoc Netw.* **2015**, *28*, 1–16. [CrossRef]

31. Li, J.; Liu, D. DPSO-based clustering routing algorithm for energy harvesting wireless sensor networks. In Proceedings of the 2015 International Conference on Wireless Communications & Signal Processing (WCSP), Nanjing, China, 15–17 October 2015; pp. 1–5.

32. Babayo, A.A.; Anisi, M.H.; Ali, I. A review on energy management schemes in energy harvesting wireless sensor networks. *Renew. Sustain. Energy Rev.* **2017**, *76*, 1176–1184. [CrossRef]

33. Sharma, V.; Mukherji, U.; Joseph, V.; Gupta, S. Optimal energy management policies for energy harvesting sensor nodes. *IEEE Trans. Wirel. Commun.* **2010**, *9*, 1326–1336. [CrossRef]

34. Liu, R.-S.; Sinha, P.; Koksal, C.E. Joint energy management and resource allocation in rechargeable sensor networks. In Proceedings of the 2010 Proceedings IEEE INFOCOM, San Diego, CA, USA, 14–19 March 2010; pp. 1–9.

35. Seyedi, A.; Sikdar, B. Energy efficient transmission strategies for body sensor networks with energy harvesting. *IEEE Trans. Commun.* **2010**, *58*, 2116–2126. [CrossRef]

36. Medepally, B.; Mehta, N.B.; Murthy, C.R. Implications of energy profile and storage on energy harvesting sensor link performance. In Proceedings of the GLOBECOM 2009—2009 Global Telecommunications Conference, Honolulu, HI, USA, 30 November–4 December 2009; pp. 1–6.

37. Koulali, M.-A.; Kobbane, A.; El Koutbi, M.; Tembine, H.; Ben-Othman, J. Dynamic power control for energy harvesting wireless multimedia sensor networks. *EURASIP J. Wirel. Commun. Netw.* **2012**, *2012*, 158. [CrossRef]

38. Li, Y.; Zhang, F.; Quevedo, D.E.; Lau, V.; Dey, S.; Shi, L. Power control of an energy harvesting sensor for remote state estimation. *IEEE Trans. Autom. Control* **2017**, *62*, 277–290. [CrossRef]

39. Amirnavaei, F.; Dong, M. Online power control optimization for wireless transmission with energy harvesting and storage. *IEEE Trans. Wirel. Commun.* **2016**, *15*, 4888–4901. [CrossRef]

40. Li, C.; Ye, M.; Chen, G.; Wu, J. An energy-efficient unequal clustering mechanism for wireless sensor networks. In Proceedings of the IEEE International Conference on Mobile Adhoc and Sensor Systems Conference, Washington, DC, USA, 7 November 2005; pp. 8–604.

41. Khan, A.U.R.; Madani, S.A.; Hayat, K.; Khan, S.U. Clustering-based power-controlled routing for mobile wireless sensor networks. *Int. J. Commun. Syst.* **2012**, *25*, 529–542. [CrossRef]

42. Kwon, T.; Gerla, M. Clustering with power control. In Proceedings of the MILCOM 1999, IEEE Military Communications (Cat. No.99CH36341), Atlantic City, NJ, USA, 31 October–3 November 1999; pp. 1424–1428.

43. Guo, S.-J.; Zheng, J.; Qu, Y.-G.; Zhao, B.-H.; Pan, Q.-K. Clustering and multi-hop routing with power control in wireless sensor networks. *J. China Univ. Posts Telecommun.* **2007**, *14*, 49–57. [CrossRef]

44. Rappaport, T.S. *Wireless Communications: Principles and Practice*; Prentice Hall PTR: Upper Saddle River, NJ, USA, 1996; Volume 2.

45. NESSCAP Inc. Nesscap Ultracapacitor Products. Available online: www.nesscap.com (accessed on 27 April 2017).

46. Younis, O.; Fahmy, S. HEED: A hybrid, energy-efficient, distributed clustering approach for ad hoc sensor networks. *IEEE Trans. Mob. Comput.* **2004**, *3*, 366–379. [CrossRef]

47. Heinzelman, W.B. Application-Specific Protocol Architectures for Wireless Networks. Doctoral Dissertation, Massachusetts Institute of Technology, Cambridge, MA, USA, 2000.

Article

A Genetic Approach to Solve the Emergent Charging Scheduling Problem Using Multiple Charging Vehicles for Wireless Rechargeable Sensor Networks

Rei-Heng Cheng [1], ChengJie Xu [2] and Tung-Kuang Wu [3,*]

[1] Information Engineering College, Yango University, Fuzhou 350015, China; reiheng@gmail.com
[2] Faculty of Computer and Software Engineering, Huaiyin Institute of Technology, Huai'an 223003, China; haxcj@126.com
[3] Department of Information Management, National Changhua University of Education, Changhua City 50074, Taiwan
* Correspondence: tkwu@im.ncue.edu.tw; Tel.: +886-937137456

Received: 26 November 2018; Accepted: 14 January 2019; Published: 17 January 2019

Abstract: Wireless rechargeable sensor networks (WRSNs) have gained much attention in recent years due to the rapid progress that has occurred in wireless charging technology. The charging is usually done by one or multiple mobile vehicle(s) equipped with wireless chargers moving toward sensors demanding energy replenishing. Since the loading of each sensor in a WRSN can be different, their time to energy exhaustion may also be varied. Under some circumstances, sensors may deplete their energy quickly and need to be charged urgently. Appropriate scheduling of available mobile charger(s) so that all sensors in need of recharge can be served in time is thus essential to ensure sustainable operation of the entire network, which unfortunately has been proven to be an NP-hard problem (Non-deterministic Polynomial-time hard). Two essential criteria that need to be considered concurrently in such a problem are time (the sensor's deadline for recharge) and distance (from charger to the sensor demands recharge). Previous works use a static combination of these two parameters in determining charging order, which may fail to meet all the sensors' charging requirements in a dynamically changing network. Genetic algorithms, which have long been considered a powerful tool for solving the scheduling problems, have also been proposed to address the charging route scheduling issue. However, previous genetic-based approaches considered only one charging vehicle scenario that may be more suitable for a smaller WRSN. With the availability of multiple mobile chargers, not only may more areas be covered, but also the network lifetime can be sustained for longer. However, efficiently allocating charging tasks to multiple charging vehicles would be an even more complex problem. In this work, a genetic approach, which includes novel designs in chromosome structure, selection, cross-over and mutation operations, supporting multiple charging vehicles is proposed. Two unique features are incorporated into the proposed algorithm to improve its scheduling effectiveness and performance, which include (1) inclusion of *EDF* (Earliest Deadline First) and *NJF* (Nearest Job First) scheduling outcomes into the initial chromosomes, and (2) clustering neighboring sensors demand recharge and then assigning sensors in a group to the same mobile charger. By including *EDF* and *NJF* scheduling outcomes into the first genetic population, we guarantee both time and distance factors are taken into account, and the weightings of the two would be decided dynamically through the genetic process to reflect various network traffic conditions. In addition, with the extra clustering step, the movement of each charger may be confined to a more local area, which effectively reduces the travelling distance, and thus the energy consumption, of the chargers in a multiple-charger environment. Extensive simulations and results show that the proposed algorithm indeed derives feasible charge scheduling for multiple chargers to keep the sensors/network in operation, and at the same time minimize the overall moving distance of the mobile chargers.

Energies **2019**, *12*, 287

Keywords: wireless rechargeable sensor networks; multiple mobile chargers; charge scheduling; genetic algorithm; sustainable networks

---

1. Introduction

In wireless sensor networks (WSN), hundreds or thousands of homogenous sensors with limited battery power are deployed in the region where information will be collected. Sensor nodes detect their proximate environment for specific information and send that information to a base station. While sensors are deployed uniformly in a wireless sensor network, sensors located in different positions bear different communication responsibilities. Accordingly, the energy consumption of sensors in the network may be heterogeneous, so some may drain their power earlier than the others and cause subsequent network failure.

Fortunately, as a result of advances in wireless charging technology [1], it is now possible to recharge the sensors through a vehicle equipped with a charger (will be referred to as mobile charger hereafter) [2–4]. WSNs that possess such a recharging capability, which are called wireless rechargeable sensor networks (WRSNs), may potentially extend their network life time. However, the scheduling of recharges needs to take the dynamic change in loading of each sensor, which usually comes in bursts when events occur, into consideration. Otherwise part of the sensors may deplete their energy before being charged, thus resulting in possible network failure. In other words, an efficient scheduling of mobile charger(s) to perform its/their duty has/have to consider varied sensor requirements. Note that, regardless of the number of mobile chargers used in a WRSN, optimal charging scheduling and its related issues have been proven to be an NP-hard problem and so far only heuristic/near-optimal solutions can be provided.

Two essential factors that are often used to determine the order of recharge are time (the sensor's deadline for recharge or energy consumption rate) and distance (from charger to the sensor demands recharge) [5,6]. The two factors have usually been combined together through some pre-determined weights in previous works. For example, in reference [5], both factors are considered equally, while reference [6] argues that a 100% (time, in terms of energy consumption rate, which is not constant as reference [5] assumes) and 80% (distance) split would be more favorable. However, as we have stated earlier, in a dynamic changing sensor network environment, it is questionable whether any static combination of the two parameters will meet all the sensors' charging requirements.

Accordingly, in this work, we propose a genetic approach to solve the emergent charging scheduling problem. The proposed algorithm is unique in adopting the following two strategies. First, the genetic optimization process incorporates the Earliest Deadline First (*EDF*) and Nearest Job First (*NJF*) scheduling outcomes into the initial chromosome population, which guarantees that the results of the genetic optimization would be no worse than the *EDF* and *NJF* results. More importantly, it also implies that both the time and distance (encoded in chromosomes based on *EDF* and *NJF* generated outcomes) factors in charge scheduling are taken into consideration, while the genetic process is used in some way to derive the weightings of the two. Secondly, the proposed algorithm may assign the charging tasks to single or multiple chargers to meet the in-time charging limitations. In the case of a multiple-charger arrangement, a clustering step is applied prior to every chromosome re-generation procedure. The clustering process proceeds by grouping neighboring sensors that demand recharges so that each group may be assigned to one of the available chargers, which thus confines the movement of each mobile charger in a more local region. The advantages of the above two strategies, as will be shown later in the simulation results, are improving the successful sensor recharge rate and greatly shortening the overall moving distance of mobile chargers. In other words, the charging need of sensors can be addressed, while reducing the energy consumption of mobile chargers. In addition, the proposed algorithm is generic as it makes no assumption regarding what target power level (i.e., charging deadline) sensors may issue charging requests and poses no restriction to both a charging vehicle's moving speed and a sensor's charging rate. However, the issue with genetic process is that scheduling may need a more powerful platform as compared to the method used in reference [5].

Overall, the objective of this research is to develop an effective genetic-based scheduling algorithm to sustain the lifetime of sensor networks in a multi-charger (include single charger) context. The novel design of chromosome structure, selection, cross-over and mutation operations, together with the afore-mentioned two strategies to fulfill the above goal will be presented and implemented. Extensive simulations with parameters setting that is as realistic as possible will be conducted to evaluate the performance of the proposed genetic approach, the effectiveness of incorporating one or both of the strategies, and the effects of the number of mobile chargers. Performance comparisons will also be made among ours and the other similar implementations. Note that a preliminary idea for this work appeared in reference [7].

The rest of the paper is organized as follows. In the related work section, we review previous works in the field of rechargeable wireless sensor networks. The proposed genetic approach to solve the charging scheduling problem is provided in the following section. Extensive simulation results of the proposed approach will be described and discussed in the simulation results and discussions section. Finally, we conclude this paper and list possible future research directions in the final section.

2. Related Works

Wireless sensor networks have received a lot of attention and have been widely used in collecting data in various fields in the past two decades [8]. However, energy drain due to the limited battery power of the sensor nodes has been one of the major issues that results in eventual network failure. However, the energy of sensors can be replenished by collecting ambient energy, such as solar power, vibration energy, and so on. Nevertheless, the energy that can be harvested is unstable and unpredictable. Fortunately, the feasibility of wireless power transfer through magnetic resonance, as demonstrated by Kurs et al. [1], makes sustainable wireless sensor networks possible. Accordingly, a WSN that possesses such a battery recharge capability, the Wireless Rechargeable Sensor Network (WRSN) has emerged to become one of the popular areas in current WSNs research [9–11].

Some research proposed to use static charging devices to charge the sensors [12,13]. The required number of charging devices may be increased proportionally to the area of the wireless sensor network, which induces more cost. Moreover, due to the limited range of wireless power transmission, the static charger solution may not be as energy efficient as compared to the mobile charger solutions [14].

In terms of mobile charger solutions, single or multiple charging vehicle(s) may be used. The former may work well in a small scale environment [15]. Yet, in a larger network, one mobile charger may not be feasible, considering the limited amount of energy it can carry. However, in both cases, one of the most essential issues is to schedule the mobile charger(s) so that it/they can charge whichever sensor that is running out of energy in time to keep the network operating.

L. Fu et al. [16] proposed an energy synchronized mobile charging (ESync) protocol, which simultaneously reduces both of the charger's travel distance and the charge waiting time of sensor nodes. Their algorithm operates by constructing a set of nested TSP (Travelling Salesman Problem) tours, which is an NP-hard problem, based on the sensor nodes' energy consumption rate. Sensor nodes are grouped according to their power consumption rate and charger is scheduled to visit nodes belonging to the group in the order of descending consumption rate. However, the possible issue with this technique would be that scheduling has to be redone once the power consumption rate of the sensors changes, which happens quite often when new events spotted in the network.

L. Xie et al. [17] showed that WSN can remain operational forever through periodic wireless energy transfer. They assumed that the node's energy consumption is fixed. After careful analysis, the near optimal charging schedule is obtained. The same team [18] also proposed to exploit the multi-node wireless energy transfer technology provided by magnetic resonant coupling. Their algorithm partitions a two-dimensional plane into adjacent hexagonal cells, and the charging vehicle visits these cells and charges multiple sensor nodes from the center of a cell. They then pursue a formal optimization framework by jointly optimizing the traveling path, flow routing and charging time. The result was a provably near-optimal solution for any desired level of accuracy. Both techniques mentioned above work on finding the (near) optimal traveling paths for mobile charger so that it may visit and charge sensor nodes periodically. In other words, the issue of on-demand recharging requirements of sensors was not addressed.

Y. Shu et al. [19] proposed to control the velocity of mobile chargers in order to reduce the traveling time of chargers so that effective time spent on sensors charging can be increased. They first proved that velocity control of mobile charger in the specified problem to be NP-hard, and then developed a heuristic solution with a provable upper bound. Although the work does not relate directly to the recharge scheduling problem, yet it indicates a potential factor to be considered in future research.

M. Pan et. al. [20] proposed a heuristic algorithm to solve the charging problem. In their method, the charging vehicle not only charges the sensors, but also collects data from some sensors acting as intermediate sinks and then brings back the information to the base station. The goal is to reduce the propagation of data packets among sensor nodes, which is the primary source of energy consumption. This appears to be a good idea, since reducing the energy consumption of sensors relieves the burden of chargers, which may deserve further attention in our future study. However, on-demand recharging requirement of sensors was still not addressed in their work.

Jie Jia et al. [21] used a heuristic algorithm to locate sensors that need charging, followed by a genetic approach to find a shorter moving sequence. The proposed strategy used only one mobile charger, but the authors did not discuss how to fulfill the dynamic requirements of sensors. A.H. Mohamed [22] also adopted a genetic algorithm to the charging vehicle scheduling problem. They used a cluster structure to decrease the complexity of a genetic algorithm. That is, their algorithm finds a better moving path among cluster heads only. The mobile charger charged the cluster heads, which then charged their members in the cluster hierarchy via energy propagation. The advantage of such a hierarchical design would thus be reducing the traveling distance of the mobile charger. However, the possibility of using multiple charging vehicles and dynamic charging requirements was not included in their research scope.

H. Dai et al. [15] proved that the optimal quantity of the charging vehicles in the wireless sensor network is an NP-hard problem, and they use an approximation approach to obtain a better solution. The authors proposed a heuristic approach to get the minimal number of charging vehicles to keep the residual energy of sensors above some working threshold. In their work, the energy consumption of each sensor is assumed to be uniform and constant, which under many circumstances may not be true. In other words, the diversity of sensor requirements should be taken into consideration in charging scheduling. Moreover, the efficient use of the charging vehicles to address the dynamic charging requests is not included in their research.

According to the aforementioned prior research, with either one charger or multiple chargers, the optimal charging scheduling and its related issues, have either been proven to be an NP-hard problem [15,16,18] or only heuristic/near-optimal solutions can be provided [18,20–22]. Accordingly, in this paper, a genetic approach like Jie Jia et al. [21] and A.H. Mohamed [22] is adopted. However, the scheduling problem will be addressed in the context of multiple mobile chargers and the dynamic sensors charging requirement will also be taken into consideration. In addition, meeting the recharging requirements of sensors while reducing the travelling distance is also one of the major issues that was addressed [16,19], which will be a major point in our proposed solution as well.

3. Genetic Approach for Multiple Mobile Chargers Scheduling

In this section, the proposed genetic approach for scheduling multiple mobile chargers in WRSN is presented, followed by the context and assumptions of the proposed approach, and the algorithm details.

First of all, assume that there are a total of N_{car} mobile chargers and the positions of sensors are already known to the base station (*BS*). In addition, a sensor node will send a recharge request to the *BS* when its battery power drops to some predefined threshold, which means the *BS* is also aware of the residuals of the sensor, i.e., its remaining work time (*RWT*). All sensor nodes, together with their remaining work time and positions, that demand charging are kept in the *charging_list*. With the above information, the *BS* may determine when to initiate a charge scheduling and determine the charging routes for the mobile chargers. The goal is to minimize the moving distance of the chargers while meeting the requirement that all sensors are recharged in time if possible. Specifically, when the *BS* finds the "start of charge scheduling" condition (as described later) occurs, it initiates the scheduling procedure (a genetic algorithm as described later in this section) and then assigns missions to the mobile chargers accordingly.

3.1. Conditions for Start of Charge Scheduling

The preliminary requirement to meet the "start of charge scheduling" condition is the availability of mobile charger(s). In other words, there must be at least one (out of N_{car}) charger available at *BS*'s disposal. The number of available chargers is denoted as $N_{car\_available}$, $0 \leq N_{car\_available} \leq N_{car}$. In case $N_{car\_available} \geq 1$, either one of the following two scenarios: (1) the number of sensor nodes demanding charging reaches the capacity of recharging (Q) that currently available chargers can provide, or (2) the minimum recharging deadline time, *min_RDT*, beyond that some sensor nodes may die if a charging mission does not start, which will trigger the initiation of a charge scheduling procedure.

The capacity of charging, Q, is the total number of sensor nodes that can be charged given a specific number of available chargers. It may be estimated according to the total battery power required by the sensors and the battery power may be provided by the available charging vehicles. To simplify the matter, let us assume that each charging vehicle is capable of charging an equal number of sensor nodes, N_{charge}. In case there are $N_{car\_available}$ chargers, Q can be expressed by Equation (1).

$$Q = N_{car\_available} \times N_{charge} \tag{1}$$

The minimum recharging deadline time (*min_RDT*), can be computed as follows. First, sort the charging tasks demanded by sensor nodes in the *charging_list* according to the Earliest Deadline First (*EDF*) order [23] and distribute these tasks evenly to all available mobile chargers. The arrival time, AT_{S_i}, of each charger to each assigned sensor node, S_i, can be calculated given the facts that visiting order, distance to sensor node and moving speed of mobile charger are known. The estimated recharging deadline time RDT_{S_i} for sensor node S_i, can be derived by computing the difference between the arrival time (of mobile charger) and its remaining work time, as depicted in Equation (2).

$$RDT_{S_i} = AT_{S_i} - LT_{S_i} \tag{2}$$

To deriving the minimum recharging deadline time (*min_RDT*) is then equivalent to finding the minimum among all the RDT_{S_i}, $S_i \in charging\_list$

3.2. The Proposed Genetic Charge Scheduling Algorithm

The proposed genetic scheduling algorithm starts by randomly generating initial population P. After an iteration (and every subsequent iteration) of the genetic process, the best $\alpha\%$ chromosomes in the population are preserved for the next generation. Also, $\beta\%$ chromosomes of the new population are newly and randomly generated as done in the initial phase. The remaining chromosomes of the new population $(1-\alpha\%-\beta\%)$ are generated by using the crossover function from two randomly selected chromosomes of previous generation, among which some might be mutated with γ probability. The genetic process would proceed until it reaches a pre determined maximum iteration count (*max_Iteration*) or the fitness function has not been further improved for t consecutive iterations.

The detailed genetic algorithm is listed in Algorithm 1. Note that Algorithm 1 would be performed n times with parameter *Number_of_car* iterates from 1 to n, where n equals to the number of charging vehicles available at the moment when the algorithm starts. The goal is finding the number of charging vehicles (k, $1 \leq k \leq n$) that achieves the best recharge scheduling.

Algorithm 1. Genetic algorithm for recharging scheduling

Input: *Number_of_car, charging_list*
Output: the best chromosome for recharging scheduling using *Number_of_car* vehicles
1   Generating initial population P.
2   Select the first chromosome from P to be the *best_one* chromosome
3   *fitness_unchanged_count* = 0
4   **for** $i = 0$ to *max_Iteration* **do**
5   Calculating fitness value of each individual in P
6   Sorting P ascendingly according to the values of individuals
7   Select the first chromosome from P to be the *new_best_one* chromosome
8   **if** *best_one* is equal to *new_best_one*
9   *fitness_unchanged_count* ++
10   **else**
11   Set *best_one* to be the *new_best_one* chromosome
12   *fitness_unchanged_count* = 0
13   **end**
14   **if** *fitness_unchanged_count* > t
15   Return the *best_one* chromosome
16   **end**
17   Set population *next_P* to {}
18   Add first α% individuals in P to next_P
19   Generate and add β% new individuals to *next_P* by using Algorithm 2
20   **for** $j = 0$ to *Population_size*$\cdot(1-\alpha\%-\beta\%)$ **do**
21   Select two parents p_1 and p_2 from P by using Algorithm 3
22   Cross p_1 with p_2 by using Algorithm 4, and produce child p'
23   Cross p_2 with p_1 by using Algorithm 4, and produce child p''
24   Choose the better one from p' and p'' to be p_{child}
25   Mutate p_{child} by using Algorithm 5 with probability γ
26   Add p_{child} into *next_P*
27   **next** j
28   **next** i
29   Return the first chromosome of P

In the following, the design and generation of chromosomes (Algorithm 2), fitness function, selection (Algorithm 3), crossover (Algorithm 4), and mutation (Algorithm 5) operations that are involved in the afore-mentioned genetic scheduling algorithm will be presented in more detail.

3.2.1. Chromosome Design and Generation

Assume that there are m sensors that need to be recharged in the entire wireless sensor network. In the case of single mobile charger, a chromosome may be designed to contain m genes, with each gene assigned the serial number of the sensor. In other words, a chromosome and a gene represent the list of sensors demand charging and the sensors ID in that list, respectively. The position of gene in a chromosome stands for the charging order of a sensor. Figure 1 depicts a possible gene combination of a chromosome in the case of seven sensors requesting a recharge. In this specific case, sensor 1 would be charged first, followed by sensors 5, 3, 2, 7, 6, and 4.

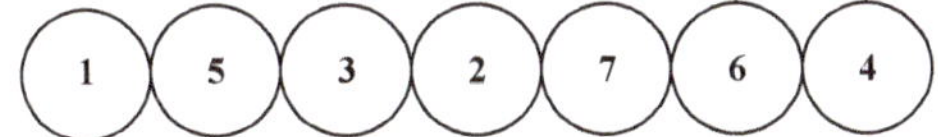

Figure 1. A sample chromosome.

To allocate the charging tasks to multiple mobile chargers, the assignment of sensors to the charging vehicles should also be included in the chromosome. Accordingly, each gene now contains not only the serial number of a sensor, but also the serial number of the charger that is responsible for recharging of that specific sensor. Figure 2 shows a scenario with 3 chargers and 7 sensors waiting to be recharged. The charging order is {1, 5, 3, 2, 7, 6, 4} and chargers 1, 2, and 3 are assigned to charge sensors {1, 2, 7}, {5, 6}, and {3, 4}, respectively. In that case, the chromosome encoding is depicted in Figure 2.

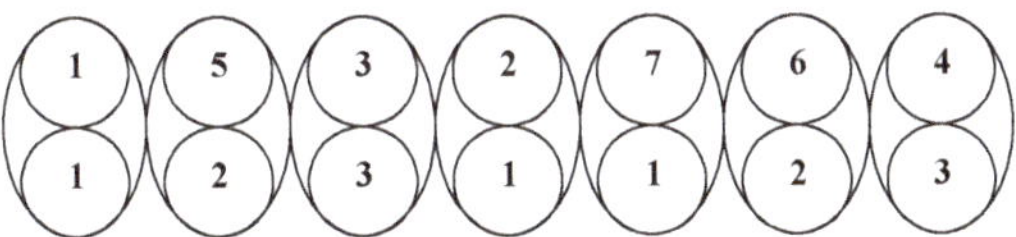

Figure 2. A sample chromosome for solving the charge scheduling problem with multiple chargers.

Algorithm 2 shows the detailed process of chromosome generation. Note that, assigning adjacent sensors demand recharging to the same charger may be an effective way to reduce the energy cost of moving chargers back and forth among sensors located apart from each other for their missions. To explore and incorporate this idea into our proposed method, we adopt a notion of clustering that is similar to the k-center algorithm. To be more specific, in case there are k available chargers and m to-be-charged sensor nodes, this process may divide these sensors into k groups based on their locality and assign them to the k available chargers. Taking the scenario depicted in Figure 3 as an example, the clustering process proceeds as follows: (1) Calculate the angle, θ_i ($1 \leq i \leq m$), of the ray originating from BS to each sensor node i (as the θ shown in Figure 3a, $-\pi < \theta \leq \pi$). (2) Sort θ_i in ascending order and derive the sensor nodes list {7, 6, 5, 4, 3, 2, 1, 8}, which may be considered as a circular queue (3) Pick randomly a sensor from the list as the starting node and assigned sensors evenly to each available chargers. For example, if the second node, 6, is selected, the assignment would go like: {6, 5, 4}→charger 1, {3, 2, 1}→charger 2, and {8, 7}→charger 3, as shown in Figure 3b.

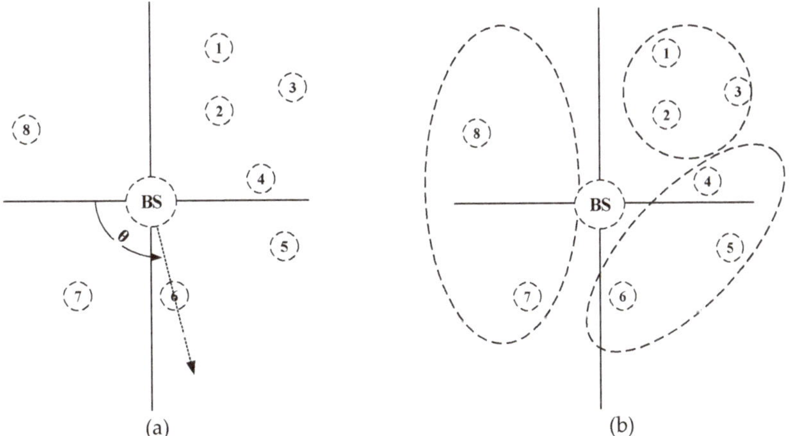

(a)   (b)

Figure 3. A scenario depicting how the clustering process in Algorithm 2 works. (**a**) Calculation of the angle of the ray originating from BS to the sensors. (**b**) Grouping and assigning sensors to be charged to mobile chargers.

Algorithm 2. Algorithm for generating chromosome instances

Input: *Number_of_available_car, charging_list*
Output: a randomly generated chromosome Chrom
1   Sort sensors in the *charging_list* in ascending order by the angle of the ray originating from BS to the sensors
     to be charged and derive new *charging _list* in angle sequence.
2   Set Chrom = {}
3   N_{charge_list} = size of *charging_list*
4   $N_{avg} = \left\lfloor N_{charging\_list} / num\_of\_available\_car \right\rfloor$
5   **for** $i = 1$ to $N_{charging\_list}$
6   Set gene.*node_id* to a randomly selected sensor ID from *charging_list*
7   Mark gene.*node_id* in *charging_list* as used
8   Set *angle_seq* as the position of gene.*node_id* in *charging_list*
9   Set gene.*car_id* = $\left\lfloor angle\_seq / N_{avg} \right\rfloor$
10   Add gene to Chrom
11 **next** *i*
12 Return Chrom

3.2.2. Fitness Function

The goal of the proposed charge scheduling algorithm is to charge all the sensors before their batteries exhausted and to minimize the energy consumption of the mobile charger. From the point of view of charging vehicles, the amount of power required for a charging mission does not change no matter what the charging sequence is. Accordingly, minimizing the moving distance of the mobile charger(s) corresponds to saving its/their energy. Before going further with our discussion, a summary of the notations to be used later is listed in Table 1.

Table 1. Notations used in this work.

Symbol	Definition
$D_{base,\,p_k^1}$	The distance from the base station to the first sensor to be charged in charging path k.
$D_{p_k^i,\,p_k^{i+1}}$	The distance between the i-th and $i+1$-th sensors to be charged in charging path k.
$\tau_{arrive}^{p_k^i}$	The time of the charging vehicle arrives at the i-th sensor of charging path k.
$\tau_{charge}^{p_k^i}$	The time needed to charge the i-th sensor in the charging path to the target power level
$Speed_{car}$	The moving speed of charging vehicles
$Speed_{charge}$	The charging speed of battery
$R_{target}^{p_k^i}$	The target power level of i-th sensor in the charging path k
$R_{residual}^{p_k^i}$	The current power level of i-th sensor in the charging path k
$RWT_{p_k^i}$	The remaining working time of i-th sensor in the charging path k
$sendtime_{p_k^i}$	The time stamp of i-th sensor in charging path k that sends charging request
$deadldine_{p_k^i}$	The deadline of i-th sensor in charging path k to be charged
$Distance_{car}^k$	The moving distance of k-th charging vehicle
$Distance_{total}$	The sum of moving distances of all charging vehicles
p_k^i	The i-th sensor in charging path k (for k-th charging vehicle)
$PathLength_k$	The length of charging path k

In case of using a single charger, the fitness function is designed to be the moving distance of the charger. Since the charging vehicle departs from the base station for charging, and returns to the base station after completing its tasks, the fitness function for some charging path p_1: $p_1^1, p_1^2, p_1^3, \ldots, p_1^n$ containing n sensors is defined as follows.

The objective is to minimize

$$f(p_1) = Distance_{car}^1, \tag{3}$$

$$Distance_{car}^1 = D_{base,p_1^1} + \sum_{i=1}^{n} D_{p_1^i, p_1^{i+1}} + D_{p_1^n, base}, \tag{4}$$

We have

$$\tau_{arrive}^{p_1^i} = \tau_{arrive}^{p_1^{i-1}} + \tau_{charge}^{p_1^i} \text{ for } 2 \leq i \leq n, \tag{5}$$

where

$$\tau_{arrive}^{p_1^1} = time_{current} + \frac{D_{base,p_1^1}}{Speed_{car}}, \tag{6}$$

and

$$\tau_{charge}^{p_1^i} = \frac{R_{target}^{p_1^i} - R_{residual}^{p_1^i}}{Speed_{charge}} \text{ for } 1 \leq i \leq n, \tag{7}$$

Since each sensor sends its charging request to the base station with the estimated remaining working time $RWT_{p_1^i}$, the base station can get the *deadldine*$_{p_1^i}$ by summing *sendtime*$_{p_1^i}$ and $RWT_{p_1^i}$. For every node i in charging path p_1, we have

$$\tau_{arrive}^{p_1^i} \leq deadldine_{p_1^i}, \tag{8}$$

While there are m charging vehicles available with m charging paths p_k: $p_k^1, p_k^2, p_k^3, \ldots, p_k^{n_k} (1 \leq k \leq m)$, the fitness function has to take into consideration the sum of all charging vehicles' moving distance. Similar to Equation (4), the $Distance_c^{ark}$ is defined below.

$$Distance_{car}^k = D_{base,p_k^1} + \sum_{i=1}^{n_k - 1} D_{p_k^i, p_k^{i+1}} + D_{p_k^{n_k}, base}, \tag{9}$$

Fitness function for a chromosome x is modified as follows.

$$f(x) = \sum_{k=1}^{m} Distance_{car}^k, \tag{10}$$

Equations (5)–(7) are modified as follows.

$$\tau_{arrive}^{p_k^i} = \tau_{arrive}^{p_k^{i-1}} + \tau_{charge}^{p_k^i} \text{ for } 2 \leq i \leq n_k, \text{ where} \tag{11}$$

$$\tau_{arrive}^{p_k^1} = time_{current} + \frac{D_{base, p_k^1}}{Speed_{car}^k} \text{ and,} \tag{12}$$

$$\tau_{charge}^{p_k^i} = \frac{R_{target}^{p_k^i} - R_{residual}^{p_k^i}}{Speed_{car}^k}, \tag{13}$$

So the problem can be formulated as a linear programming formula as follows:
Objective function:

$$Minimize \sum_{k=1}^{m} Distance_{car}^k, \text{ subject to } \tau_{arrive}^{p_k^i} \leq deadldine_{p_k^i} \text{ for } 1 \leq i \leq n_k \text{ and } 1 \leq k \leq m. \tag{14}$$

Note, only the scenarios where the charging vehicles move at constant speed and charge sensors at the same rate will be considered in the following discussions. But according to equations above (i.e., Equations (12) and (13)), there is no restriction on both the charging vehicle's moving speed and the sensor's charging rate. In addition, sensors can issue requests at different target power levels and provide different charging deadline. As a result, the fitness function as designed and presented above can potentially deal with dynamic sensors' charging requirements with multiple mobile chargers, which may be an interesting problem to be addressed in the future.

In addition, the charging time should also be included in the fitness function. A schedule that cannot complete the charging tasks is given a larger fitness value so that this schedule has lower possibility of being selected. On the other hand, for charging schedules that can complete charging tasks in time should be given scores according to the times they needed for these tasks. Finally, the fitness function is defined as follows.

$$f(x) = a \times \tau_{\text{overtime}} + b \times \tau_{\text{total}} + c \times Distance_{total}, \tag{15}$$

$$\tau_{\text{overtime}} = \sum\nolimits_{k=1}^{m} \sum\nolimits_{i=1}^{PathLength_k} \left(deadldine_{p_k^i} - \tau_{\text{arrive}}^{p_k^i} \right), \tag{16}$$

$$\tau_{\text{total}} = \max_{1 \le k \le m} \left(\tau_{arrive}^{p_k^{PathLength_k}} + \frac{D_{p_k^{PathLength_k},base}}{Speed_{car}^k} \right), \tag{17}$$

$$Distance_{total} = \sum_{k=1}^{m} Distance_{car}^k, \tag{18}$$

The constants a, b, c used in Equation (15) are predefined constant values as the weights of τ_{overtime}, τ_{total} and $Distance_{total}$, respectively. In the simulations performed in this work, a is set to a very larger number (e.g., 1,000,000) to avoid adopting charging schedules that are not able to fulfill their missions. On the other hand, b and c are both set to 1 for simplicity. In the future, it may be possible to explore various combinations of these three parameters to potentially improve scheduling performance.

3.2.3. Selection

In order to ensure that better chromosomes may possess higher probabilities of being selected to generate new chromosomes, the chromosome pool is first sorted in ascending order according to the fitness values (smaller fitness value implies better chromosome, which would be placed at the forepart of pool). Assume that there are n chromosomes in the pool. Then a random number s is selected from 0 to $n^2 - 1$ and the $(n - \lfloor \sqrt{s} \rfloor)$-th chromosome would be selected. To sum up, the algorithm to select a chromosome from the chromosome pool is listed below.

Algorithm 3. Algorithm for randomly selecting a chromosome

Input: an integer number n
Output: a chosen chromosome
1   Select a uniformly distributed random number s from 0 to n^2-1
2   Choose the $(n - \lfloor \sqrt{s} \rfloor)$-th chromosome from the sorted chromosome pool to be the selection result

---

3.2.4. Crossover

Crossover proceeds by first randomly selecting two chromosomes C_a and C_b from the chromosome pool using the selection method mentioned above (Algorithm 3). A cutting point k is then randomly generated. Let $C[i]$ represent for the gene at position i of chromosome C, and $C[i, j]$ represents a subsequence of genes of chromosome C from the position i to j. Then the content of the crossover chromosome C_c is composed of $C_a[0, k - 1]$ and $C_b[k, n - 1]$. However, since the contents of $C_a[0, k - 1]$ and $C_b[k, n - 1]$ may be duplicated, the crossover steps are modified accordingly as follows.

For example, a possible crossover process, as depicted in Algorithm 4, of two chromosomes, C_a and C_b, with each containing 5 genes and with cutting point k being set to 3 is shown in Figure 4. As can be seen, the crossover result, C_c, receives its first 3 (=k) and last 2 genes from C_a and C_b, respectively. However, since the two genes from C_b contain sensor IDs that have overlapped with the ones from C_a, they would be marked with -1 and later been replaced with ones that are not shown in genes from C_a, which would be sensor ID 4 and 5.

Algorithm 4. Algorithm for chromosomes' crossover

Input: Two chromosomes C_a and C_b.
Output: A new chromosome C_c
1   Randomly select an integer k, where $1 \leq k < |C_a| - 1$.
2   Set $C_c[0, k-1] = C_a[0, k-1]$.
3   Set $C_c[i] = \begin{cases} -1, & if\ C_b[i] \in C_c[0,\ k-1] \\ C_b[i], & otherwise \end{cases}$
    where $k \leq i \leq n - 1$
4   Find the elements from C_b that have not yet been placed in C_c, and fill them sequentially to the positions
    marked with -1 in C_c.

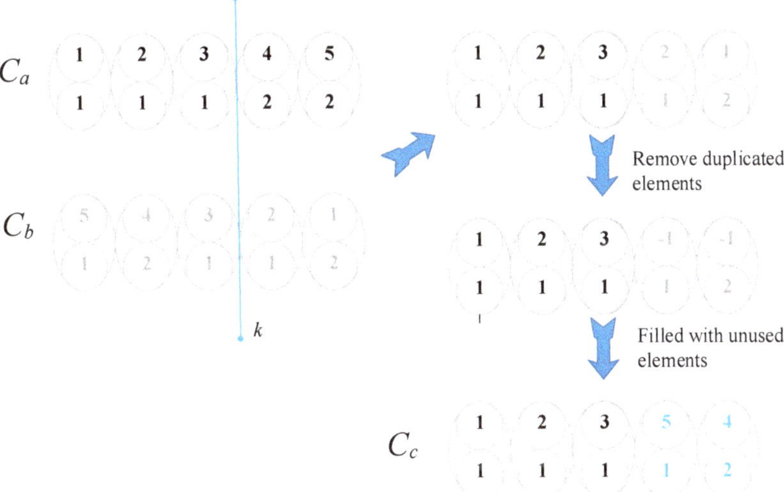

Figure 4. An illustration of the crossover operation.

3.2.5. Mutation

As a chromosome should contain each sensor to be charged just once, a sensor ID in a chromosome cannot just change to another one randomly. Thus, the mutation operation is designed to swap the IDs of two randomly selected genes in a chromosome. The algorithm to mutate a chromosome C_m is presented in Algorithm 5 below. Figure 5 is an illustration of the mutation process depicted in Algorithm 5.

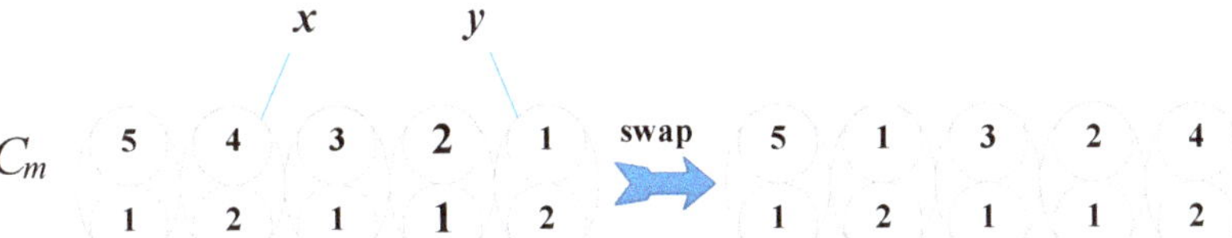

Figure 5. An illustration of the mutation operation.

Algorithm 5. Algorithm for mutating a chromosome

Input: chromosome C_m
Output: mutated chromosome C_m
1   Select two distinct number x and y randomly, $0 \leq x, y \leq n - 1$.
2   Swap the ID values of $C_m[x]$ and $C_m[y]$.

---

4. Simulation Results and Discussions

The simulation program was coded with Visual Studio C# and executed on a personal computer with an Intel i7-8700 CPU running Microsoft Windows 10. The simulation environment is specified by a 1000×1000 meters' rectangular region with 1000 sensors uniformly and randomly deployed. The base station is located at the center of the area. Mobile chargers are equipped with independent and unlimited power source for their movement. Also assume that a gradient based routing tree [24] has been built already. Packets are sent from nodes to the base station hop by hop. Based on the free space model proposed in reference [25], to transmit l bits over d distance would cost $E_{Tx}(l, d)$ and to receive l bits would cost $E_{Rx}(l)$ as shown below.

$$E_{Tx}(l,\ d) = lE_{elec} + l\epsilon_{fs}d^2$$

$$E_{Rx}(l) = lE_{elec}$$

$$E_{elec} = 50\frac{nJ}{bit}$$

$$\epsilon_{fs} = \frac{10pJ}{bit}/m^2$$

The traffic load of each node is assigned a random value uniformly distributed within range $[0, x/n]$, where n is the total number of nodes. In other words, in every second, each sensor node has a probability in between 0 and x/n in generating a packet. Whenever a node finds that its remaining energy is lower than the preset percentage (E_{th}), it needs to send the charging request packet to the base station. Both the request packet and the data packet are fixed at a size of 10 kB in the following simulations. Packet transmission is set to 1/100 s for one hop delivery. The content of a charging request packet includes node *id*, energy consumption rate R_c, residual energy E_c, and demand time $T_{current}$. Energy consumption rate R_c and Remaining Work Time (*RWT*) are calculated using the following equations.

$$R_c = \frac{E_{full} - E_c}{T_{current} - T_{full}},$$

where E_{full} and T_{full} are battery capacity and time that the sensor was last charged.

$$RWT = E_c/R_c$$

Unless otherwise specified, the parameters used in simulations are summarized in Table 2. All the results are the average of 100 simulations. Three performance metrics, as listed below, are used for evaluating the simulation results.

- Percentage or number of sensors been charged—the percentage of sensors can be successfully charged. This is the primary goal of the proposed algorithm: the more sensors been charged successfully, the longer the network life time would be expected.
- Moving distance—the average moving distance of the charging vehicles in charging a sensor. It is calculated by dividing the total moving distance of charging vehicles by the total number of sensors been charged. The shorter the average moving distance, the lower the cost in charging the sensors.
- Percentage of data packets been delivered—the percentage of all transmitted data packets that are successfully delivered to their destinations.

Note, as shown in Table 3, most of our parameter settings are based on those used by Tsoumanis et al. [26], and three other research articles [5,6,17]. In some cases, the values of our parameters were chosen to be even harsher than theirs. Examples of this were the network traffic, battery power of chargers and sensor nodes' initial energy power. Also, according to our assessment (through simulation), the energy consumption rate of the free space model is also higher than the other models in our scenario, and thus puts more pressure on the network. Finally, other than the assumption that mobile chargers have an independent and unlimited power source for their movement, all the other parameter settings are chosen so that they are as close as possible to a realistic scenario.

Table 2. Simulation parameter settings.

Parameters	Values
Network size	$1000 \times 1000\ \mathrm{m}^2$
Number of nodes	1000
Energy capacity of node	500 J
Energy capacity of mobile charger	10000 J
Traveling speed of mobile charger	5 m/s
Energy consumption of moving mobile charger	0
Charging rate	5 J/s
Initial residual of nodes	5%~25% ([0..20%] + 5%)
Residual of sensor triggering charging request	10%
Energy consumption model	Free space model
Network traffic (Probability of a node generating a packet per second)	[0, 1/100], [0, 1/10]
Transmission range of nodes	60 m
Packet length	10 kB
Parameters t, α, β, and γ used in Algorithm 1	20, 10, 10, 0.2
Population size (of GA procedure)	200
max_Iteration (used in Algorithm 1)	200

Table 3. Parameter settings used in other methods.

	[5]	[6]	[17]	[26]
Battery full	100 kJ	500 J	10.8 kJ	1
Network size	$100 \times 100\ \mathrm{m}^2$	$100 \times 100\ \mathrm{m}^2$	$1000 \times 1000\ \mathrm{m}^2$	1×1
Number of nodes	100	20–100	50, 100	1000
Transmission range	—	—	—	0.06
WCV traveling speed	4 m/s	3 m/s	5 m/s	—
WCV energy	—	50,000 J	—	—
WCV moving consumption	—	10 J/m	—	—
Packet length	—	—	—	—
Charging speed	—	—	5 J/s	—
Network traffic(for each node)	—	—	[1,10 kb/s]	[0, 1/1000]
Node initial energy	full	—	—	—
Energy consumption model	0.002 kJ/s	TB-Powercast	Multi-path	μr^3, μ=1,3,5
Request threshold	30%	225 J (45%)	540 J (5%)	10%, 50%, 90%

We have designed and conducted five simulations, with the results shown in the following sub-sections.

4.1. The Impacts of Genetic Algorithm Parameters

The first simulation is designed to assess the two important parameters, population size and maximal number of iterations, of a genetic procedure. To speed up the simulation, the simulation time is limited to 10,000 seconds with network traffic set at $[0, 100/n]$, while the other parameters are the same as shown in Table 2. In these simulations, the values of both the population size and maximal number of iterations are set in between 200 and 800. The following figures show the results of all the three performance metrics listed above plus the average computation time (the average time required for completing a scheduling task on the simulation personal computer) by varying the population size and maximal number of iterations. Figure 6 reveals that increasing the value of population size only slightly improves the percentage of sensors being charged and packets received, and the average moving distance. On the other hand, the increase in maximal number of iterations not only has no apparent impact on the performance, but also greatly prolongs the computation process. This may imply that the genetic algorithm usually converges before it reaches the maximal iteration counts. Thus, in the following simulations, it may be adequate to set the population size and the maximal number of iteration to 200 and 200, respectively.

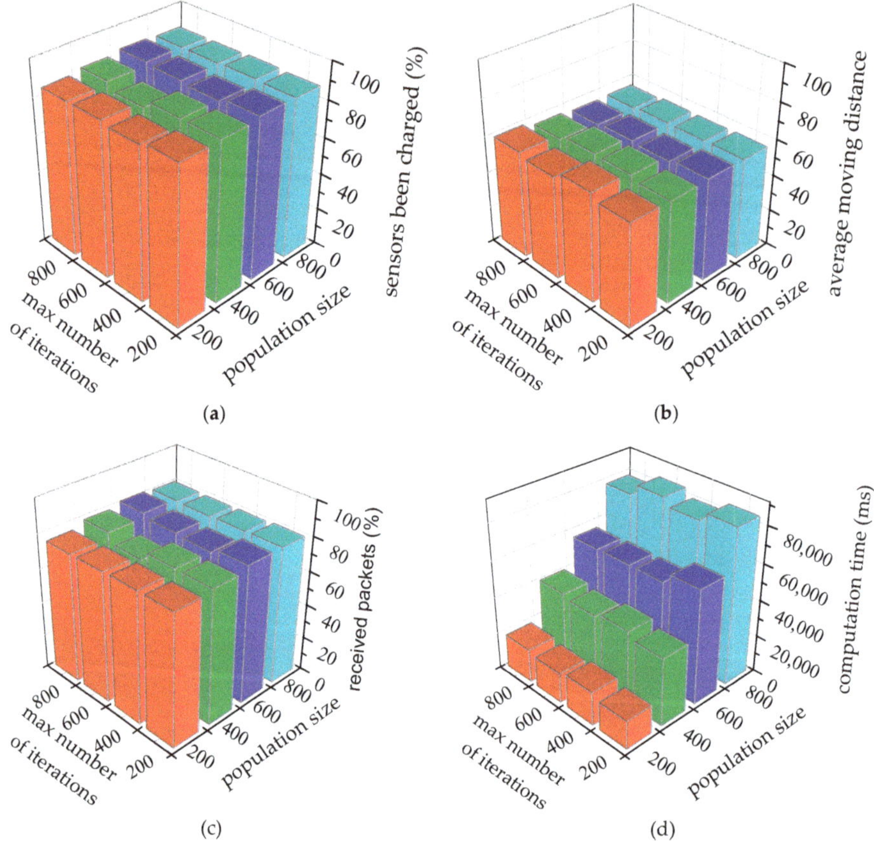

Figure 6. The performance metrics of the proposed genetic algorithm when varying population sizes and maximal number of iterations. (**a**) percentage of sensors been charged, (**b**) average moving distance, and (**c**) percentage of packets received, and (**d**) average computation time, vs population sizes and maximal number of iterations. Note that simulation time and network traffic are set at 10,000 s and [0, 100/n], while the other parameters, except population size and maximum number of iterations, are the same as shown in Table 2.

4.2. The Impacts of Vehicle Parameters

The second simulation is to investigate the effects of charging vehicle's moving speed and sensor charging speed, which determine the cost (in time) of charging tasks. Intuitively, the smaller the time cost, the easier it is to find a feasible solution to complete the charging tasks. As expected, Figure 7a,c show that with a higher vehicle speed and charging rate, it is easier to get charging tasks done, which again improves the packets received rate. Figure 7b shows that when charging tasks can be mostly completed (vehicle speed $\geq$ 5 m/s and charging rate $\geq$ 5 J/s), both vehicle speed and charging rate do not have much impact on the average moving distance (of charging vehicles). Accordingly, the following simulations will set the speed of vehicle and battery charging rate to 5 m/s and 5 J/s, respectively.

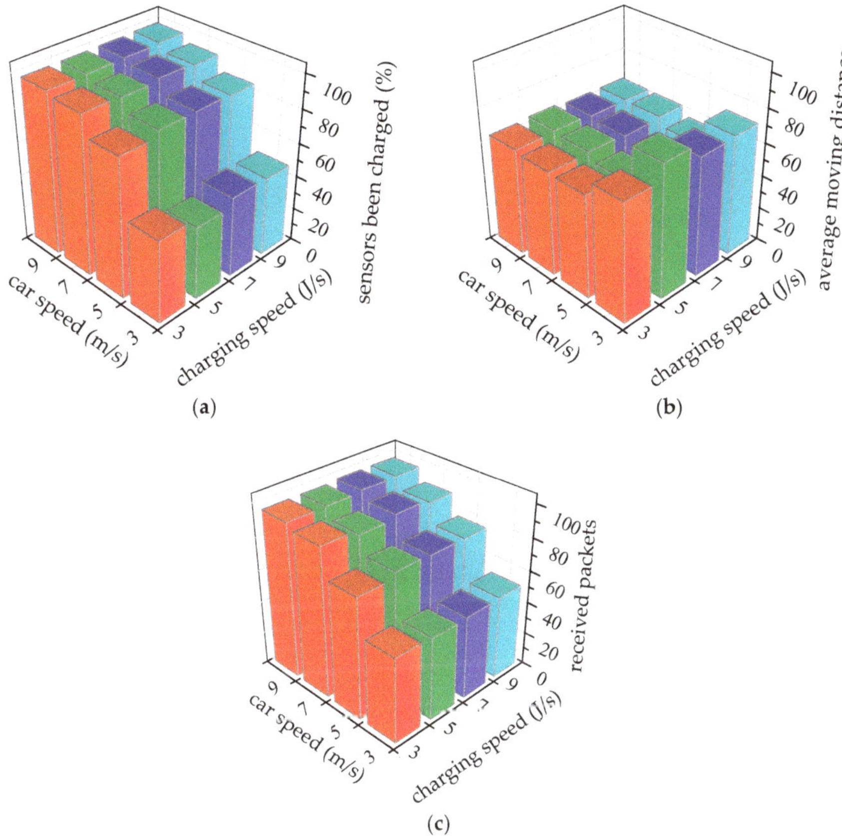

Figure 7. The performance metrics under different vehicle moving speed and battery charging rate. (**a**) percentage of sensors been charged, (**b**) average moving distance, and (**c**) percentage of packets received, vs moving speed of charging vehicle and charging rate of sensor. Note that simulation time and network traffic are set at 10,000 s and $[0, 100/n]$, while the other parameters, except for vehicle moving speed and sensor charging speed are the same as shown in Table 2.

4.3. The Impacts of Clustering and Incorporation of EDF/NJF Outcomes to Initial Chromosomes

In the following simulation, performance of variations of our proposed genetic scheduling approach is examined. Five possible variations include: implementations with clustering step only (denoted by Clustering), with clustering step plus incorporating *EDF* and/or *NJF* scheduling outcomes into the initial chromosome (denoted by *EDF*-Clustering, *NJF*-Clustering, and EDF + NJF-Clustering), and with none of the above (denote Random, meaning no clustering is performed and all initial chromosomes are generated randomly).

From the results shown in Figure 8, it is obvious that, when comparing to pure genetic algorithm (Random), the full implementation of all additional steps (*EDF* + *NJF* + Clustering, which includes clustering and *EDF* + *NJF*) of our proposed algorithm performs significantly better in all three performance metrics. It is also interesting to see the impacts of each extra steps. First of all, clustering alone does not have any physical effect when there is only one charger, which is within our expectations. Clustering only starts to have moderate performance margin over the random genetic implementation in multi-charger scenarios. Second, NJF-Clustering appears to perform better (in terms of successful recharging and data packet delivery rate) than *EDF*-Clustering in cases with less available chargers. However, as the number of chargers increases (≥ 3), the performances of the two start going the opposite directions. It appears that when there are more chargers at our disposal, the weighting in "time" is more essential than the "distance" factor. Note, *NJF*-Clustering does still perform better in terms of average moving distance, which is reasonable as it takes "distance" as its first priority. Third, although *EDF + NJF* + Clustering implementation consistently achieves less in average moving distance of chargers, it seems to be the only one that this metric increases as the number of cars. The reason for this may be due to the sometimes conflicting circumstances between the "time" and "distance" factors, and the compromise made by the genetic procedure causes the increase in average moving distance.

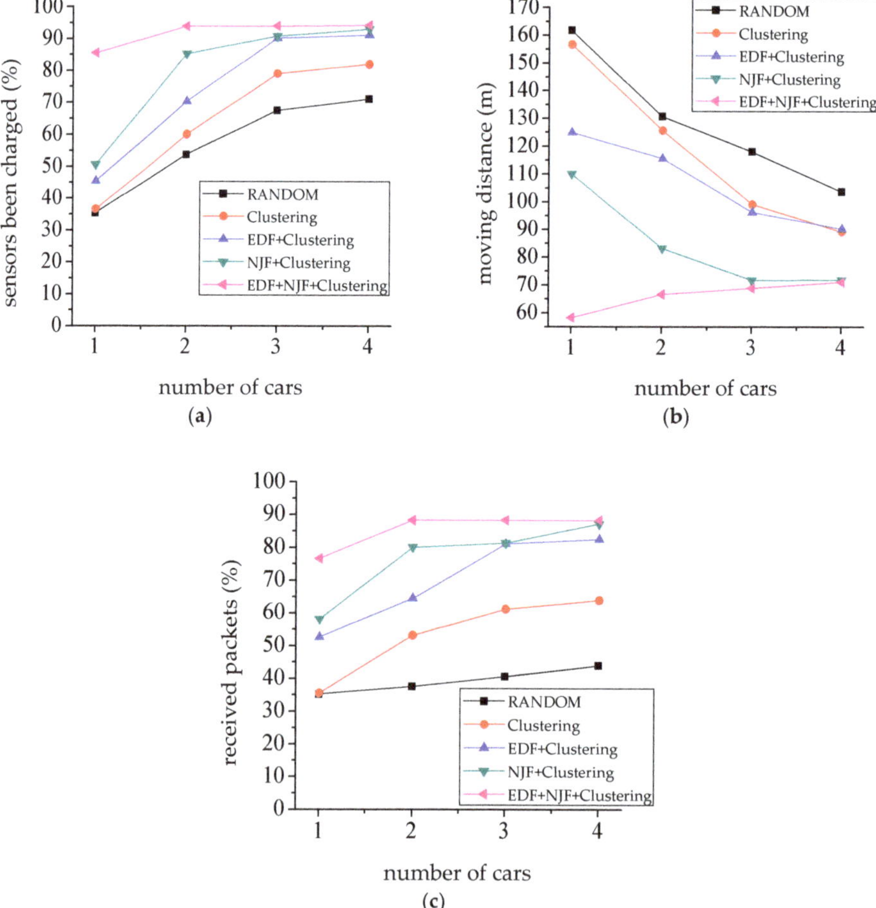

Figure 8. Performance comparisons among variations of our proposed algorithm in terms of: (**a**) percentage of sensors being charged successfully, (**b**) average moving distance of charging vehicles, and (**c**) percentage of successful data packets delivery. Note that simulation time and network traffic are set at 10,000 s and [0, 100/n], while the other parameters are the same as shown in Table 2.

4.4. The Impacts of Varying Number of Chargers

In this simulation, we try to investigate the effects of increasing the number of chargers under light ([0..10/n]) and heavy ([0..100/n]) network traffic scenarios. Note, the full implementation of our algorithm (i.e., the EDF + NJF + Clustering implementation) is adopted in this and the following simulations. Also, the simulation time is extended to 1,000,000 s to explore the comparatively long-term effects of our algorithm.

As shown in Table 4, we can see that under light traffic conditions, an increase in the number of chargers available does not improve much in terms of both a successful sensor charging rate and data packet delivery rate. On the other hand, it does show a positive impact as network traffic becomes heavier, especially in terms of data packet delivery rate. Note that less than 100% of the percentage of sensors that have been charged may be due to the strict setting in the initial residual of nodes (in-between 5% and 25%) and the hot spot effect around the base station, which result in the early death of nodes without having any chance of being recharged. In addition, we may also observe the results and realize how the failure of even a small percentage of sensors may affect the final successful data delivery. For example, in a scenario with heavy network traffic and one available charger, a less than 4% rate of recharge misses may cause drops in packet delivery by more than a quarter.

When it comes to moving distance, more chargers mean more add-on costs induced by moving these chargers back to the base station for energy replenishment, and thus increase the average moving distance of charging vehicles. Overall, having more chargers is definitely helpful in prolonging a sensor network's operations. However, the average moving distance of chargers in light traffic conditions all appear to be longer. It seems that there are less recharge requests expected in that case. As a result, each scheduled mission of a charger turns out to service less sensor nodes, and thus increases the average moving distance per sensor charged. Also, due to the same afore-mentioned reasons (the strict setting in initial residual of nodes and the hot spot effect around the base station), the chargers (initially located at the base station) usually tend to service nearby sensor nodes in the beginning of the simulation. As the simulation proceeds over time, the effect would be less obvious. That is why we see results with significantly lower average moving distances (less than 100 m) from previous simulations when their simulation time was set to 10,000, instead of 1,000,000 s in this simulation and the following ones.

Table 4. Simulation results with varying numbers of chargers.

Number of cars	Sensor Been Charged (%)		Moving Distance (m)		Packet Delivery Rate (%)	
	[0..10/n]	[0..100/n]	[0..10/n]	[0..100/n]	[0..10/n]	[0..100/n]
1	99.810%	96.041%	199.9	218.6	99.998%	72.188%
2	100%	97.080%	256.3	221.9	99.999%	74.596%
3	99.810%	98.008%	263.7	232.2	99.999%	80.925%
4	99.683%	99.084%	262.6	240.9	99.998%	94.895%

(Note, simulation time and network traffic are set at 1,000,000 s, while the other parameters are the same as shown in Table 2.).

4.5. Comparison of Scheduling Performance with Other Mrthods

In the final simulation, we compare scheduling performance between ours and the afore-mentioned algorithm [5], which uses static combination of "time" and "distance" with the same weight. The algorithm was called TemporAl and Distantial Priority charging scheduling algorithm (denoted by TADP). EDF scheduling outcomes are also included and used as the baseline for comparison. The simulations are done under both light ([0..10/n]) and heavy ([0..100/n]) network traffic scenarios with one charger. The results, as shown in Table 5, indicate that our algorithm outperforms the other two implementations in all three performance metrics under both network loading conditions.

Table 5. Performance comparison between ours and the other methods.

Method	Sensor Been Charged (%)		Moving Distance (m)		Packet Delivery Rate (%)	
	[0..10/n]	[0..100/n]	[0..10/n]	[0..100/n]	[0..10/n]	[0..100/n]
EDF	94.479%	87.473%	346.3	292.2	89.272%	70.094%
TADP	89.270%	87.640%	335.1	289.4	78.603%	70.201%
Ours	99.810%	96.041%	199.9	218.6	99.998%	72.188%

(Note, simulation time and network traffic are set at 1,000,000 s, while the other parameters are the same as shown in Table 2).

One thing that deserves noting is that TADP performs worse than EDF in the lighter traffic condition, and only does slightly better when traffic is heavier. It appears that given "time" and "distance" the same weight as TADP may result in a scenario with a sensor node closer to the charger but with an less urgent need to be recharged first. In the meantime, another sensor that needs immediate attention yet located a few hops away may just drain its last bit of power. On the other hand, EDF would have made a completely different decision, which may just have enough time to save both sensors from dying in a less busy network environment. In fact, the lower successful recharge percentage / data delivery rate and shorter average moving distance of TADP (as compared to EDF) imply that this would very likely be the case. However, when network traffic becomes heavier, "time"-first or "time and distance"-balanced do not seem to have much difference. Either way, the results indicate that charge scheduling that depends on putting on some pre-determined and static weights upon the "time" and "distance" may not be a good idea. This, along with a much better performer in both light and heavy traffic conditions, demonstrates that the dynamic nature of our proposed recharge scheduling algorithm may well be a better answer to the constantly changing sensor network environment.

5. Conclusions

In this paper, a genetic approach to solve the emergent charging scheduling problem with multiple charging vehicles for rechargeable wireless sensor networks is proposed. The multiple charging vehicles scheduling problem is nicely transformed to a genetically evolving process through a novel design of chromosome structure, selection, cross-over and mutation operations. The major goal is to ensure a sustainable sensor network, while at the same time minimizing the traveling distance of the mobile chargers.

In terms of the number of mobile chargers used in a WRSN, there can be one or multiple ones available. However, with either one charger or multiple chargers, the optimal charging scheduling and its related issues have been proven to be an NP-hard problem and only heuristic/near-optimal solutions can be provided. Our proposed genetic-based scheduling algorithm is distinct and different from the others in the following three aspects. First of all, the proposed algorithm incorporates the *EDF* (Earlier Deadline First) and *NJF* (Nearest Job First) scheduling algorithm generated outcomes into the initial populations (aside from the other randomly generated ones), which guarantees that the results of our genetic optimization would be no worse than the *EDF* and *NJF* results. Second, we adopt an extra step where clusters sensors would be charged to k groups based on their locality and assigned to k available chargers in newly generated populations of each generation. This step ensures sensor nodes that are closer been assigned to the same charger later on, which then effectively reduces the moving distance of chargers. Finally, the fitness function we developed imposes no restriction on the charging vehicle's moving speed and the sensor's charging rate. Sensors are also allowed to issue requests at different target power levels and provide different charging deadlines. These facts imply that the fitness function, and thus our genetic approach, can be used to deal with dynamic sensors' charging requirements in the multiple mobile chargers scenarios. In other words, the proposed genetic algorithm may be applied to WRSNs with heterogeneous sensors and mobile chargers.

With extensive simulations, the impacts of many parameters in the proposed genetic algorithm are revealed and investigated further so that more insights may be gained, and adjustments made accordingly. The simulations also show that the proposed approach can find efficient scheduling paths for multiple charging vehicles while making sure virtually all the sensors in need are charged in time. In addition, the idea of clustering charging requests made by adjacent sensors and assigning these requests to the same mobile charger has proven to be a very good strategy in reducing the overall moving distance of chargers, and thus saving battery power and efficiency in charging the sensors.

For future research, it would be interesting to apply the proposed genetic algorithm to WRSNs with heterogeneous sensors and mobile chargers. In addition, designing a more efficient genetic approach to deal with more sensors and more charging requirements so that the overall computation time may be reduced in order to respond promptly to more realistic and complex situations will be our other concern. Finally, our algorithms assume that mobile chargers always return to the base station for recharge until its/their next missions. We also assume there would always be some available charger(s) ready for deployment. In reality, the recharging of the chargers themselves [27,28] (includes battery for charging sensor nodes and battery for their movement, and whether or not the base station is the only source for providing energy to the chargers) is another issue that deserves further exploration, which will also be our future research topic.

Author Contributions: Conceptualization, R.-H.C.; Formal analysis, R.-H.C.; Methodology, R.-H.C., C.X. and T.-K.W.; Resources, C.X.; Validation, C.X. and T.-K.W.; Writing—original draft, R.-H.C. and T.-K.W.; Writing—review & editing, T.-K.W.

Funding: This research was funded by Yango University, Fuzhou 350015 China.

Conflicts of Interest: The authors declare no conflict of interest.

References

1. Kurs, A.; Karalis, A.; Moffatt, R.; Joannopoulos, J.D.; Fisher, P.; Soljačić, M. Wireless power transfer via strongly coupled magnetic resonances. *Science* **2017**, *317*, 83–86. [CrossRef] [PubMed]
2. Kumar, A.R.; Gayathri, H.R.; Gowda, B.R.; Yashwanth, B. WiTricity: Wireless Power Transfer by Non-radiative Method. *Int. J. Eng. Trends Technol. (IJETT)* **2014**, *11*, 290–295. [CrossRef]
3. Barman, S.D.; Reza, A.W.; Kumar, N.; Karim, M.E., Munir, A.B. Wireless powering by magnetic resonant coupling: Recent trends in wireless power transfer system and its applications. *Renew. Sustain. Energy Rev.* **2015**, *51*, 1525–1552. [CrossRef]
4. Hui, S.Y.R. Magnetic Resonance for Wireless Power Transfer [A Look Back]. *IEEE Power Electron. Mag.* **2016**, *3*, 14–31. [CrossRef]
5. Lin, C.; Wang, Z.; Han, D. TADP: Enabling temporal and distantial priority scheduling for on-demand charging architecture in wireless rechargeable sensor networks. *J. Syst. Archit.* **2016**, *70*, 26–38. [CrossRef]
6. Ping, Z.; Yiwen, Z.; Shuaihua, M. RCSS: A Real-Time On-Demand Charging Scheduling Scheme for Wireless Rechargeable Sensor Networks. *Sensors* **2018**, *18*, 1601.
7. Cheng, R.H. A Genetic Approach to Solve the Charging Scheduling Problem for Wireless Rechargeable Sensor Networks. In Proceedings of the International Conference on Recent Innovations in Engineering and Technology (ICRIET), Prague, Czech Republic, 9–10 July 2015.
8. Akyildiz, I.F.; Su, W.; Sankarasubramaniam, Y.; Cayirci, E. Wireless sensor networks: A survey. *Comput. Netw.* **2002**, *38*, 393–422. [CrossRef]
9. Akbari, S. Energy harvesting for wireless sensor networks review. In Proceedings of the 2014 Federated Conference on Computer Science and Information Systems, Warsaw, Poland, 9–10 September 2014; pp. 987–992.
10. Oueis, J.; Stanica, R.; Valois, F. Energy Harvesting Wireless Sensor Networks: From Characterization to Duty Cycle Dimensioning. In Proceedings of the 2016 IEEE 13th International Conference on Mobile Ad Hoc and Sensor Systems (MASS), Brasilia, Brazil, 10–13 October 2016; pp. 183–191.
11. Shaikh, F.K.; Zeadally, S. Energy harvesting in wireless sensor networks: A comprehensive review. *Renew. Sustain. Energy Rev.* **2016**, *55*, 1041–1054. [CrossRef]

12. He, S.B.; Chen, J.M.; Jiang, F.C.; Yau, D.K.Y.; Xing, G.L.; Sun, Y.X. Energy Provisioning in Wireless Rechargeable Sensor Networks. *IEEE Trans. Mobile Comput.* **2013**, *12*, 1931–1942. [CrossRef]

13. Liao, J.H.; Hong, C.M.; Jiang, J.R. An adaptive algorithm for charger deployment optimization in wireless rechargeable sensor networks. In Proceedings of the 2014 International Computer Symposium, Taichung, Taiwan, 12–14 December 2014; pp. 2080–2089.

14. Erol-Kantarci, M.; Mouftah, H.T. Mission-Aware Placement of RF-based Power Transmitters in Wireless Sensor Networks. In Proceedings of the 2016 IEEE Symposium on Computers and Communications (ISCC), Messina, Italy, 1–4 June 2012; pp. 12–17.

15. Dai, H.; Wu, X.; Chen, G.; Xu, L.; Lin, S. Minimizing the number of mobile chargers for large-scale wireless rechargeable sensor networks. *Comput. Commun.* **2015**, *46*, 54–65. [CrossRef]

16. Fu, L.; He, L.; Cheng, P.; Gu, Y.; Pan, J.; Chen, J. ESync: Energy Synchronized Mobile Charging in Rechargeable Wireless Sensor Networks. *IEEE Trans. Veh. Technol.* **2016**, *65*, 7415–7431. [CrossRef]

17. Xie, L.; Shi, Y.; Hou, Y.T.; Sherali, H.D. Making sensor networks immortal: An energy-renewal approach with wireless power transfer. *IEEE/ACM Trans. Netw.* **2012**, *20*, 1748–1761. [CrossRef]

18. Xie, L.; Shi, Y.; Hou, Y.T.; Lou, W.; Sherali, H.D.; Midkiff, S.F. On Renewable Sensor Networks with Wireless Energy Transfer: The Multi-Node Case. In Proceedings of the 2012 IEEE SECON, Seoul, Korea, 18–21 June 2012; pp. 10–18.

19. Shu, Y.; Yousefi, H.; Cheng, P.; Chen, J.; Gu, Y.; He, T.; Shin, K.G. Near-optimal Velocity Control for Mobile Charging in Wireless Rechargeable Sensor Networks. *IEEE Trans. Mobile Comput.* **2015**, *15*, 1699–1713. [CrossRef]

20. Pan, M.; Li, H.; Pang, Y.; Yu, R.; Lu, Z.; Li, W. Optimal energy replenishment and data collection in wireless rechargeable sensor networks. In Proceedings of the 2014 IEEE Global Communications Conference (GLOBECOM), Austin, TX, USA, 8–12 December 2014; pp. 125–130.

21. Jia, J.; Chen, J.; Deng, Y.; Wang, X.; Aghvami, A.H. Joint Power Charging and Routing in Wireless Rechargeable Sensor Networks. *Sensors* **2017**, *17*, 2290. [CrossRef] [PubMed]

22. Mohamed, A.H. Optimizing the Performance of Wireless Rechargeable Sensor Networks. *IOSR J. Comput. Eng.* **2017**, *19*, 61–69.

23. Bouakaz, A.; Gautier, T.; Talpin, J.-P. Earliest-deadline First Scheduling of Multiple Independent Dataflow Graphs. In Proceedings of the 2014 IEEE International Workshop on Signal Processing Systems, Belfast, UK, 20–22 October 2014; pp. 20–22.

24. Intanagonwiwat, C.; Govindan, R.; Estrin, D.; Heidemann, J.; Silva, F. Directed diffusion for wireless sensor networking. *IEEE/ACM Trans. Netw.* **2003**, *11*, 2–16. [CrossRef]

25. Heinzelman, W.B.; Chandrakasan, A.P.; Balakrishnan, H. An application-specific protocol architecture for wireless microsensor networks. *IEEE Trans. Wirel. Commun.* **2002**, *1*, 660–670. [CrossRef]

26. Tsoumanis, G.; Oikonomou, K.; Aïssa, S.; Stavrakakis, I. A recharging distance analysis for wireless sensor networks. *Ad Hoc Netw.* **2018**, *75*, 80–86. [CrossRef]

27. Gupta, V.; Reddy, S.; Kumar, R.; Panigrahi, B.K. Multi-Aggregator Collaborative Electric Vehicle Charge Scheduling (CEVCS) Under Variable Energy Purchase and EV Cancellation Events. *IEEE Trans. Ind. Inform.* **2017**, *14*, 2894–2902. [CrossRef]

28. Wei, Z.; Li, Y.; Zhang, Y.; Cai, L. Intelligent Parking Garage EV Charging Scheduling Considering Battery Charging Characteristic. *IEEE Trans. Ind. Electron.* **2018**, *65*, 2806–2816. [CrossRef]

Article

Localization Approach Based on Ray-Tracing Simulations and Fingerprinting Techniques for Indoor–Outdoor Scenarios

Antonio Del Corte-Valiente *, José Manuel Gómez-Pulido, Oscar Gutiérrez-Blanco and José Luis Castillo-Sequera

Department of Computer Engineering, Polytechnic School, University of Alcalá, 28871 Alcalá de Henares, Spain
* Correspondence: antonio.delcorte@uah.es; Tel.: +34-91-885-6594

Received: 7 July 2019; Accepted: 29 July 2019; Published: 31 July 2019

Abstract: The increase of the technology related to radio localization and the exponential rise in the data traffic demanded requires a large number of base stations to be installed. This increase in the base stations density also causes a sharp rise in energy consumption of cellular networks. Consequently, energy saving and cost reduction is a significant factor for network operators in the development of future localization networks. In this paper, a localization method based on ray-tracing and fingerprinting techniques is presented. Simulation tools based on high frequencies are used to characterize the channel propagation and to obtain the ray-tracing data. Moreover, the fingerprinting technique requires a costly initial learning phase for cell fingerprint generation (radio-map). To estimate the localization of mobile stations, this paper compares power levels and delay between rays as cost function with different distance metrics. The experimental results show that greater accuracy can be obtained in the location process using the delay between rays as a cost function and the Mahalanobis distance as a metric instead of traditional methods based on power levels and the Euclidean distance. The proposed method appears well suited for localization systems applied to indoor and outdoor scenarios and avoids large and costly measurement campaigns.

Keywords: distributed networks; fingerprinting; localization; ray-tracing; wireless sensors

1. Introduction

Localization is an important aspect of GPS-alternative positioning systems and has considerable importance in general communications areas [1]. The development of new radio access standards prompted the exploration of new techniques to improve the location accuracy, which is based on the signals available from the wireless devices that comprise these standards [2–6]. In this communication, the problem of the indoor location, which is based on the signals available in the wireless devices that comprise Wi-Fi and Wi-Max networks within the broadband wireless systems, is presented [7–10]. The location process employs a fingerprinting technique that operates with the relationship between the levels of power and/or relative delays between signals due to the multipath by reflections effect [11]. By comparing other techniques, such as the angle of arrival (AOA) or time of arrival (TOA), which presents several limitations due to multipath effects and non-line of sight (N-LOS), implementation of the fingerprinting technique is relatively easy [12]. In traditional indoor location systems based on Wi-Fi networks, the Euclidean distance between the radio frequency (RF) power levels of the received signals and the RF levels stored in a database (radio-map) is employed as a metric in the location process [13]. Due to an increase in new broadband networks (UWB), for example, Wi-Max, the exploration of new techniques is necessary to improve the location accuracy using alternative detection methods based on the relative delays of available signals in the observing points, which are provided by these emergent technologies [14,15]. In this study, we proposed a novel fingerprinting technique using the relative ray delays due to the ray-tracing effects as fingerprints instead of RSSI. Greater accuracy will be obtained by this technique.

The remainder of this paper is organized as follows: In Section 2, a review of the localization techniques is presented. Section 3 describes the algorithms and techniques that have been developed and introduces the ray-tracing theory. In Section 4, simulated scenarios are introduced, and the corresponding results are presented. Section 5 discusses the challenges and open issues of the localization in real-world situations. Conclusions are presented in Section 6.

2. Review of Techniques for Localization

As our research includes a study of localization algorithms in distributed wireless networks that are applied to hybrid scenarios, describing the techniques that can be employed is important. In this section, we briefly review the state of the techniques that can be applied for localization in hybrid indoor–outdoor environments. The following techniques are described:

1. Time of arrival (ToA)
2. Time difference of arrival (TDoA)
3. Angle of arrival (AoA)
4. Received signal strength (RSS)
5. Hybrid techniques
6. Fingerprinting technique

2.1. Time of Arrival

In the technique of location by time of arrival (ToA), the distance between a mobile and an antenna is directly proportional to the propagation time [16]. To determine the position in 2D, the measurements made with the ToA technique must be made with respect to at least three reference points. By geometric calculations (technique known as lateration), the points of intersection in the ToA circles are obtained, the propagation time in one direction is measured and the distance between the transmitter and the receiving unit is calculated [17].

2.2. Time Difference of Arrival

The idea of the time difference of arrival (TDoA) technique is to determine the relative position of the mobile by examining the time difference in which the different signals reach the receiver instead of obtaining the absolute time of arrival. For each TDoA measurement, the transmitter must resolve a hyperboloid between the two units of measurement [18].

2.3. Angle of Arrival

The angle of arrival (AoA) technique, which is also referred to as the direction of arrival (DoA), uses multi-array antennas to determine the angle of arrival of the incoming signal from the mobile target via spatial diversity of the antennas [19,20].

In the AoA technique, the estimation of the target mobile can be obtained by intersecting pairs of line angles, which are each formed by a circle of radius from the base station to the target mobile [21,22].

2.4. Received Signal. Strength (RSS)

The RSS detection method is based on the loss of power that the signal experiences in the propagation medium as the power of the signal in the free space decays with the square of the distance. This method applies the received signal strength indicator (RSSI) parameter, which collects the power with which the signal from the target mobile to be located arrives at the receiving station [23–25]. The distance between the target mobile and the base stations can be calculated using the RSSI value considering the losses in the propagation.

Due to the various shadows and multipath effects in indoor environments, path-loss models are not always available. In addition, the parameters of the model directly depend on the scenario. The accuracy of this technique can be increased using the RSSI measurements in areas around the base stations or multiple measurements from several base stations [26,27].

2.5. Hybrid. Techniques

Implementation of a hybrid system that combines AoA and ToA [28,29], in which ToA forms a circle to which it cuts a straight AoA line, is possible and offers greater accuracy and a shorter response time [30]. Another possible solution is to combine ToA and TDoA. With this system, the deficiencies of the TDoA technique in NLOS environments are improved [31]. AoA and TDoA have also been combined to improve accuracy and limit multipath effects, which enables locating only with two base stations via the intersection of the straight line and the hyperbola of the AoA and TDoA measurements [32,33]. Some techniques combine measures of time and power [34].

2.6. Fingerprinting Techniques

In recent years, location-based services (LBS) have proliferated; however, the main challenge is to determine the locations of mobile terminals within a certain accuracy.

As we have indicated, traditional localization techniques based on radio frequency are subjected to strong fluctuations in a signal due to the NLOS between transmitters-receivers and other causes, such as the multipath effect, which produces variations in the information of the parameter RSSI applied as a pattern in the localization process.

To avoid these signal fluctuations, an increase in the number of access points in the scenario is possible to minimize the cases of NLOS. This idea served as the basis for introducing the fingerprinting technique, which consists of comparing the available power levels in a set of points, as fingerprints, in known positions with those obtained in the target mobiles to be located when they are excited by the same antennas. The final position of the mobile is determined by assigning the coordinates of the fingerprint of the shortest distance in the scenario.

The fingerprinting technique is a deterministic localization method that is divided into two phases:

1. Off-line phase: It constitutes the calibration phase in which a database or radio-map is generated with the power measurements obtained on the mesh fingerprinting when excited by the antennas of the environment.
2. On-line phase: it constitutes the testing phase; a significant number of mobile stations are randomly located within the coverage area of the radio-map to obtain the corresponding power measurements when excited by the same antennas.

The vector with the values of the different RSSI measurements obtained in a certain position of the mesh is referred to as the location of the fingerprinting in this point. The estimation of the location is obtained by an algorithm that computes the distance between a fingerprint and a target. In most cases, the Euclidean distance, although other distance metrics can be employed (refer to Section 3.1), between the RSS values in the target mobiles (on-line phase) and the RSS values of each fingerprint stored in the radio-map (off-line phase) is computed. The coordinates associated with the fingerprint that provides the shortest distance metric are utilized as the final position for the target mobile (Figure 1).

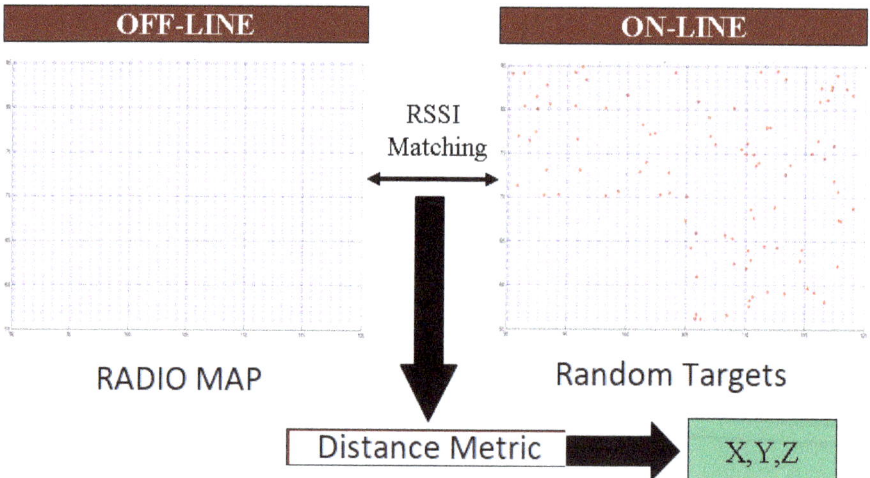

Figure 1. Fingerprinting technique: (**Left**) off-line phase; (**Right**) on-line phase.

The most common application of this method is the location in indoor environments [35–37]. In Ref. [38], the properties of the RSSI signal are analysed. In Ref. [39], a study is presented with the estimation of the expected error using this technique. In Ref. [40], a study of channel modelling is presented. In Ref. [41], an improvement of the algorithm is presented using interpolation with neighbouring fingerprints (K-NN).

For outdoor applications, we identify a classic study in Ref. [42]. Using fingerprinting outside the Wi-Max standard, a localization proposal is presented in Ref. [43]. Conversely, in Ref. [44], the fingerprinting technique is presented as a complement to the external location when the GPS signal is degraded or not available.

One of the largest problems with this technique used for outdoor locations is the large number of fingerprints that is needed in the off-line phase of the algorithm to obtain the required accuracy. In Refs. [45,46], clustering techniques are proposed to minimize this impact. In Ref. [47], hyperbolic techniques are employed to optimize the changes in the environment in the off-line phase.

As one of the main problems is the variability of the radio signal [48], learning techniques are proposed in Ref. [49] to predict the fluctuations in the signal. Similarly, Ref. [50] introduces neural networks that enable modelling of the channel.

Additionally, several studies expand the possibilities of the fingerprinting technique, which is primarily employed with Wi-Fi, to other standards and applications. In Ref. [51], a neural network on UWB is introduced. Other standards include ZIGBEE in Ref. [52] and RFID in Ref. [53]. In Ref. [54], the fingerprinting technique is combined with a Kalman filter to estimate trajectories in indoor environments.

2.7. Summary

This section presents a comparison of several figures of merit, such as, cost, precision, energy efficiency, and complexity, related with the techniques previously presented. As can be seen in Table 1, the fingerprinting algorithm combined with the hybrid detection method (in our study, power level and ray-delay) present the best results.

Table 1. Summary of localization techniques.

Algorithm	Cost	Precision	Efficiency	Complexity	Hardware
ToA	high	medium	low	large	large
TDoA	low	high	high	large	medium
AoA	high	low	medium	large	large
RSS	low	medium	medium	low	small
Hybrid	low	high	high	low	small
Fingerprinting	medium	high	high	low	small

3. Novel Localization Techniques

In this section, the localization techniques developed in this study are presented. Although these techniques are based on the fingerprinting method, instead of using the RSSI information of the received power detected in each fingerprint, information about the relative delay between two rays due to multipath effects is employed. The final objective is to increase the accuracy of a location in hybrid indoor–outdoor scenarios, where existing localization solutions do not guarantee adequate accuracy or coverage [55–57].

First, different distance metrics that are proposed to match the off-line and on-line phases of the fingerprinting technique are introduced. Second, different detection techniques that are used as cost functions are detailed. Last, additional techniques that improve the localization accuracy are presented.

3.1. Distance Metrics

A distance metric is a key component that is utilized by the fingerprinting technique in the localization process [58]. Exploration of different similarity measures is important to determine the best distance metric that minimizes the positioning average error. Four metrics have been implemented to explore improving the location accuracy. The vector of sample F denotes the data from the fingerprints (radio-map), vector of sample T denotes the data from the target mobiles, and N is the number of antennas available in the scenario.

In the Euclidean distance Equation (1), the mobile station will have more confidence in the fingerprint radio map if the distance is smaller. A more moderate approach is obtained with the Manhattan distance Equation (2) using the sum of the absolute differences rather than their squares as the total measure of dissimilarity. In the Bray-Curtis Equation (3) distance, the numerator signifies the difference, and the denominator normalizes the difference. For the Mahalanobis distance Equation (4), the $(F\text{-}T)'$ term denotes the $(F\text{-}T)$ transpose vector, and the Cov term denotes the covariance matrix, where the retrieval performance is sensitive to the scenario topology. The idea is to find the coordinates of $(x_i, y_i)_T$ that minimize the distance to $(x_i, y_i)_F$:

$$D_E(F,T) = \sqrt{\sum_{i=1}^{N}(F_i - T_i)^2} \tag{1}$$

$$D_M(F,T) = \sum_{i=1}^{N}|F_i - T_i| \tag{2}$$

$$D_{BC}(F,T) = \sum_{i=1}^{N}\frac{|F_i - T_i|}{F_i + T_i} \tag{3}$$

$$D_{Mah}(F,T) = \sqrt{(F-T)'Cov(F)^{-1}(F-T)} \tag{4}$$

3.2. Cost Function Based on RSSI

When the detection method employed in the fingerprinting technique is based on power levels, the distribution of RSSI is obtained from each fingerprint and each access point available in the scenario. Figure 2 shows a distribution of the power levels due to all fingerprints in a generic scenario. It shows the power level (in dBm units) in each fingerprint available in the radio-map. The level of power depends on the position of the fingerprint with respect to each AP, and hence, of the scenario. Lower values of power mean that a specific fingerprint is not seen by all APs. Conversely, higher values of power mean that the fingerprint is seen by most APs (an accumulative function is used to add to each fingerprint the values of power received by each AP). The map of power is unique, it is the information that identifies each fingerprint, and is used in the cost function represented by Equation (5).

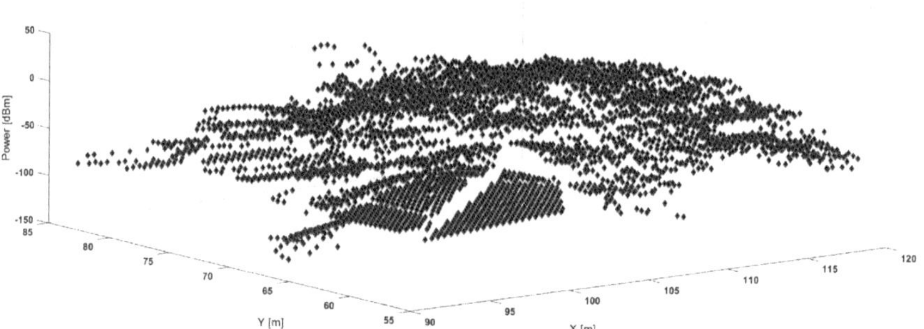

Figure 2. Distribution of power levels in the radio map. It represents the visibility of each fingerprint with respect to each AP available in the scenario, and it is the information used in the cost function based on RSSI.

For example, using the Euclidean Distance as a metric, the cost function measures the distance based on power levels (in dBm units) between each fingerprint $(RSS(x, y)_F)$ and each target mobile $(RSS(x, y)_T)$ and for each antenna (N) available in the scenario Equation (5).

$$D_E(F_i, T_i) = \sqrt{\sum_{i=1}^{N} \left((RSS(x_i, y_i)_F - RSS(x_i, y_i)_T)^2 \right)}$$

(5)

3.3. Cost Function Based on Relative Ray-Delay

When the detection method used in the fingerprinting technique is based on the relative ray-delay, the Time Delay Ray (TDR) profile is obtained from each fingerprint and each access point available in the scenario. This detection method is based on the information provided by the ray-tracing due to the different effects: direct ray, reflected ray, diffracted ray, and any combination of them. Figure 3 shows an example of a ray-delay profile obtained in one fingerprint of the scenario.

This figure represents the arrival time and the power of each ray due to the different effects. The first ray to arrive is the direct ray between the AP and the receiver, the others, that arrive more delayed, are due to reflections, diffractions, double reflections, etc.; produced by the different obstacles present in the scenario. The TDR is the delay of each ray with respect to the first or the direct ray.

If the Euclidean Distance is used as a metric, the cost function measures the distance based on the ray-tracing profile (in nanoseconds units) between each fingerprint $(TDR(x, y)_F)$ and each target mobile ($TDR(x, y)_T$), considering E as the number of ray-tracing effects and N as the number of access points (AP) in the scenario Equation (6).

$$D_E(F_i, T_i) = \sqrt{\sum_{i=1}^{N}(\sum_{j=1}^{E}\left(TDR(x_i, y_i)_F - TDR(x_i, y_i)_T)^2\right)} \tag{6}$$

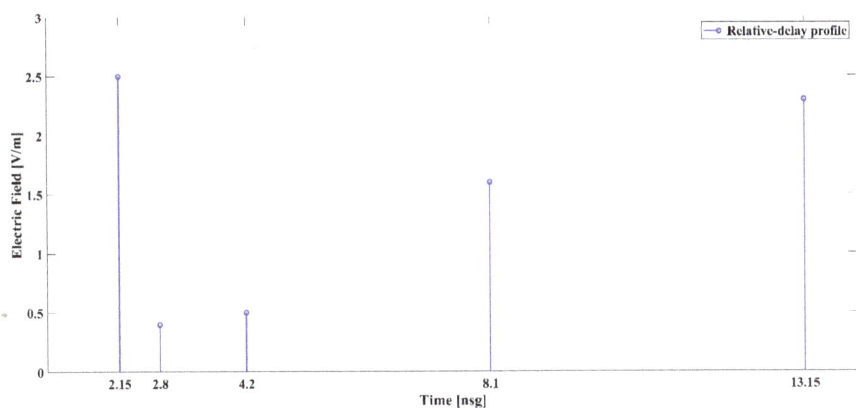

Figure 3. Relative delays and levels between rays due to ray-tracing effects.

3.4. Hybrid. Detection

An alternative detection method consists in using both the power levels and the relative delays between two rays in a hybrid manner. The μ parameter was used to consider power levels or ray-delays. By applying this method in the Euclidean distance metric (Equation (8)), if the value of $\mu = 0.25$ is already set (it was set to 0.25 by experimentation), we perform a hybrid detection between power levels and relative ray-delays. In this case, more weight is assigned to detection by ray-delays in fingerprints whose power level is at least four times below the power level of the corresponding mobile with respect to the same access point. Equation (7) can be easily extended to the remaining metrics.

$$D_E(F_i, T_i) = \mu \cdot \sqrt{\sum_{i=1}^{N}\left((RSS(x_i, y_i)_F - RSS(x_i, y_i)_T)^2\right)} + (1-\mu) \cdot \sqrt{\sum_{i=1}^{N}(\sum_{j=1}^{E}\left(TDR(x_i, y_i)_F - TDR(x_i, y_i)_T)^2\right)} \tag{7}$$

The localization technique has been implemented by a novel detection algorithm that combines the profile of the relative ray-delay obtained from ray-tracing due to multi-path effects with different distances employed as the metric.

Figure 4 shows the pseudo-code of the algorithm used for hybrid detection. The algorithm first obtains the map of effects for both targets and fingerprints. Due to the multipath effects of the ray-tracing, only fingerprints and targets with the same map of effects (reflections, diffractions, and its combinations) are considered. If the map of effects is different, then the fingerprint is not considered and the next is processed. When a fingerprint and a target have the same map of effects, and for all the antennas used in the simulations, the algorithm computes the vector of metrics from a specific target to each fingerprint. If the algorithm reaches the last fingerprint, then it finds the minimal distance between the target and all the fingerprints and uses the coordinates of that fingerprint as position of the target. Finally, the algorithm repeats the process for the rest of the targets.

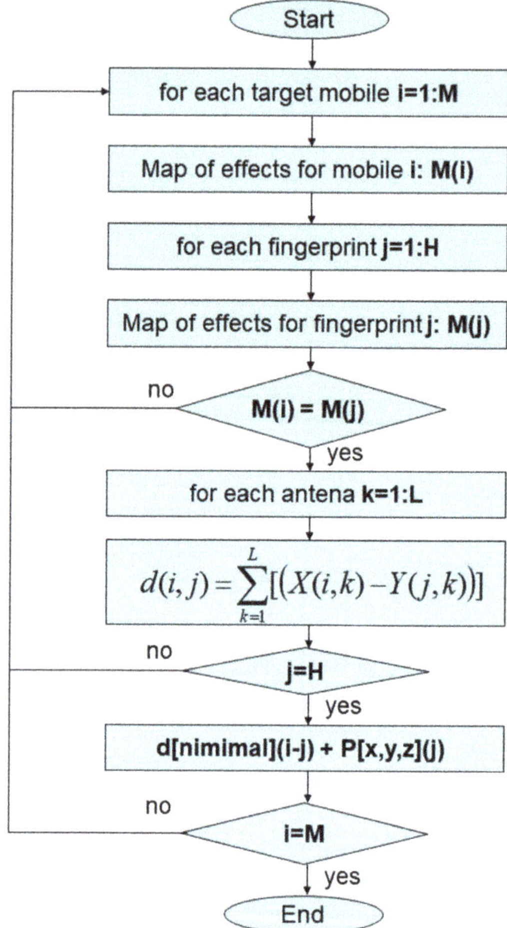

Figure 4. Pseudo-code of the localization algorithm.

3.5. Interpolation

As an interpolation technique, we propose the K-Nearest Neighbours (K-NN) algorithm that considers the nearest k neighbours, instead of a single neighbour by assigning the majority category of the set of K neighbours. The interpolation technique is applied in the on-line phase of the fingerprinting technique by seeking the k-fingerprints that are closest to the target mobile based on the error information obtained when applying the corresponding distance metric.

In addition, a pre-filter removes incoherent values from the radiomap when a specific transmitter is not seen at all (out of range or defective). In these cases, an interpolation with neighbouring values is applied.

Figure 5 shows the concept of interpolation over four random fingerprints, and Equation (8) represents the mathematical concept, where N is the number of fingerprints employed in the interpolation process, and X_j is the value of the distance between the mobile to be located and the j-fingerprint. *OldPosition* refers to the (x, y, x) coordinates of the target before the interpolation, and *NewPosition* refers to the new target coordinates after the interpolation.

$$NewPosition\ (x, y, z) = \frac{\sum_{j=1}^{N} \left(\frac{OldPosition\ (x,y,z)}{X_j} \right)}{\sum_{j=1}^{N} \left(\frac{1}{X_j} \right)} \tag{8}$$

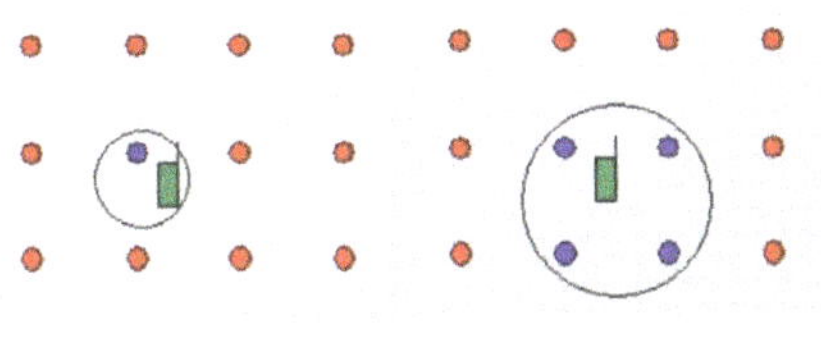

Figure 5. Interpolation technique implemented with four fingerprints.

4. Ray-Tracing Theory

Powerful electromagnetic simulation tools, which are capable of accurately reproducing the values of radioelectric signals that would be obtained in the real case, have been developed. NewFasant [59] is a tool that simulates antennas on board platforms, such as satellites and aircrafts, and radio wave propagation in urban and indoor scenarios [60–62]. This software has been developed by the Electromagnetic Computational group of the Alcala University and is based on asymptotic techniques, such as physical optics, combined with the Physical Theory of Diffraction (PO/PTD) and Geometrical Optics (GO) with the Uniform Theory of Diffraction (UTD) [63,64].

The NewFasant programme can be applied to different electromagnetic problems: study of the propagation in outdoor/indoor environments, mobile communications, radar cross-section of complex structures, electromagnetic compatibility, analysis and design of antennas, and radiation patterns of antennas on board complex structures.

Given a source point (transmitter antenna), the ray-tracing technique determines the number of rays that arise to a given observation point. To achieve this goal, possible reflections and diffractions in the environment are considered. In addition, a detailed description of the environment is important for a precise calculation of the rays involved in the propagation model. The geometrical information uses Non-Uniform Rational Bezier Splines (NURBS) surfaces. To obtain the ray-tracing paths between a source and a destination, two steps exists:

1. Obtain all points that define the ray-path (reflection and diffraction points).
2. Discard the rays occluded by any part of the geometry to perform the shadowing test.

Figure 6 shows an example of ray-tracing in a complex scenario. This area is an area of Terminal 4 of the Adolfo-Suarez Barajas Airport in Madrid. A parked airplane is placed beside a finger, an antenna is placed in front of the aircraft, and an observation point is placed in front of the antenna. The NewFasant software computes all rays generated from the antenna to the receiver.

In Figure 6, the antenna is the red dot. The observation point of black colour is the point that can be seen in front of the plane. The colour of the rays depends on the effect: yellow is the double reflection and the other colours correspond to other effects (reflection, reflection-diffraction, etc.).

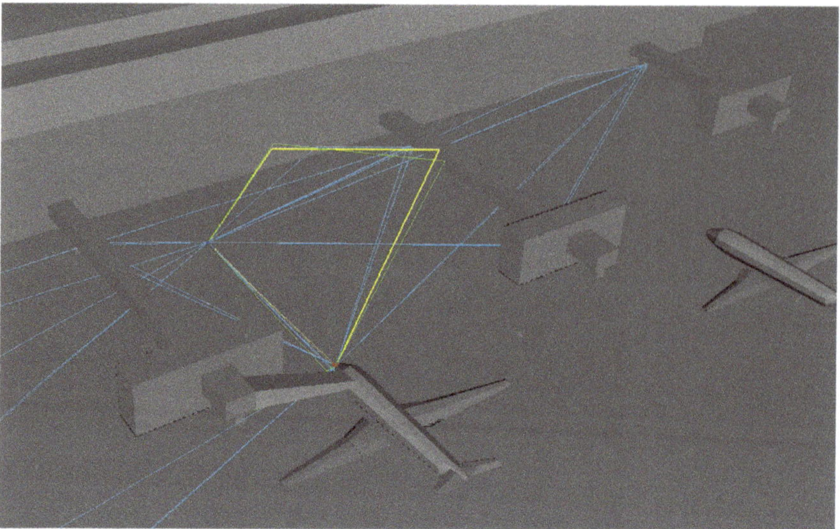

Figure 6. Ray-tracing scenario at the airport.

NewFasant provides associated information about each ray, such as the power level at the observation point (RSS), time of arrival (ToA) and direction of arrival (TDoA). Different ray-tracing effects (direct ray, reflected ray, and diffracted ray) and multiple combinations of them, such as the double reflections effect and reflection-diffraction effect, can be taken into account (Figure 7). With this information provided by NewFasant, the profile of the relative delays between rays (as shown in Figure 3) can be obtained, and the localization cost function can be computed according to expression 7.

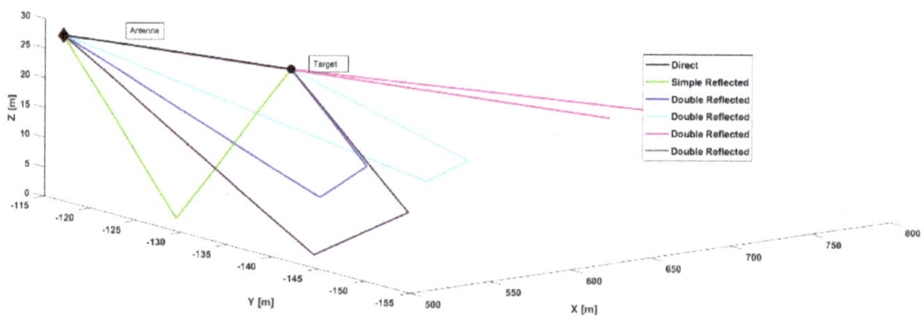

Figure 7. Different rays and effects computed by NewFasant software.

The computational cost associated with ray tracing is very high but speeding algorithms, such as Angular Z-buffer [65], can considerably reduce the computational time. This reduction enables the analysis of an excessive number of observation points in complex structures, such as an airport with parked aircrafts in a short period.

Table 2 shows the computational cost obtained for different configurations of the number of antennas and number of reflections.

Table 2. NewFasant computational cost.

Configuration	X Points	Y Points	Total Points	Computational Cost (Hours)
5 antennas 2 reflections	140	120	16,800	70
5 antennas 3 reflections	12	145	1740	104
10 antennas 2 reflections	100	50	5000	55

5. Validations and Results

The aim is to simulate various localization scenarios using the techniques described here, using powerful and precise ray tracing tools. First, we simulate typical indoor scenarios. Second, we conduct outdoor simulations in more complex environments.

The performance of any localization algorithm depends on the quality of the scenario under testing. For this reason, in each scenario, a complex CAD model of plane facets for the geometrical description of the environment was used to study the fast fading caused by multipath propagation and the interference between the different paths. Because the accuracy of the results depends on network geometry [66,67], different granularities (grid size) were used in the simulations. To evaluate the performance of each algorithm in the different tested scenarios, two statistical indicators, which will be applied to make comparisons, have been established. These indicators are the total average error (Av_{Error}), as an indicator of first order, and the typical deviation (Typ_{Dev}), as a second-order indicator. The results have been analysed in the MATLAB environment.

5.1. Indoor Scenarios

The indoor scenario that is analysed corresponds to a real section of the Polytechnic building in Madrid, Spain (Figure 8).

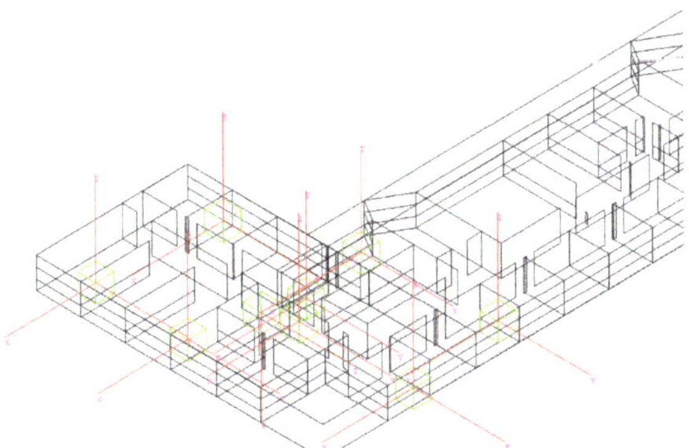

Figure 8. Indoor scenario: A representative section of the University Polytechnic of Madrid (Spain).

Four different scenarios were analysed. A total of 99 target mobiles and 9 APs were employed at two different frequencies (2.4 and 5.2 GHz):

1. Grid size of 72 × 72 m with 5184 fingerprints at 2.4 GHz
2. Grid size of 72 × 72 m with 5184 fingerprints at 5.2 GHz
3. Grid size of 36 × 36 m with 1296 fingerprints at 2.4 GHz
4. Grid size of 36 × 36 m with 1296 fingerprints at 5.2 GHz

Figure 9 shows a 2D view of the simulation area that corresponds with scenario number 3.

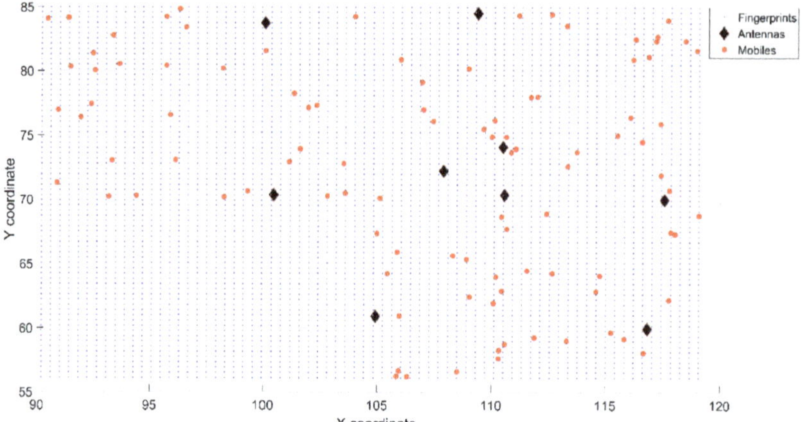

Figure 9. 2D view of the University Polytechnic of Madrid simulation area: A section of 36×36 m containing 1296 fingerprints, 99 mobiles to locate and 9 APs.

Indoor simulations are focused on the granularity of the scenario and the frequency. The goal is to evaluate the impact of the detection method (power, ray-delays and hybrid) for these variations. The Euclidean distance was employed as a metric. The results are shown in Figures 10 and 11.

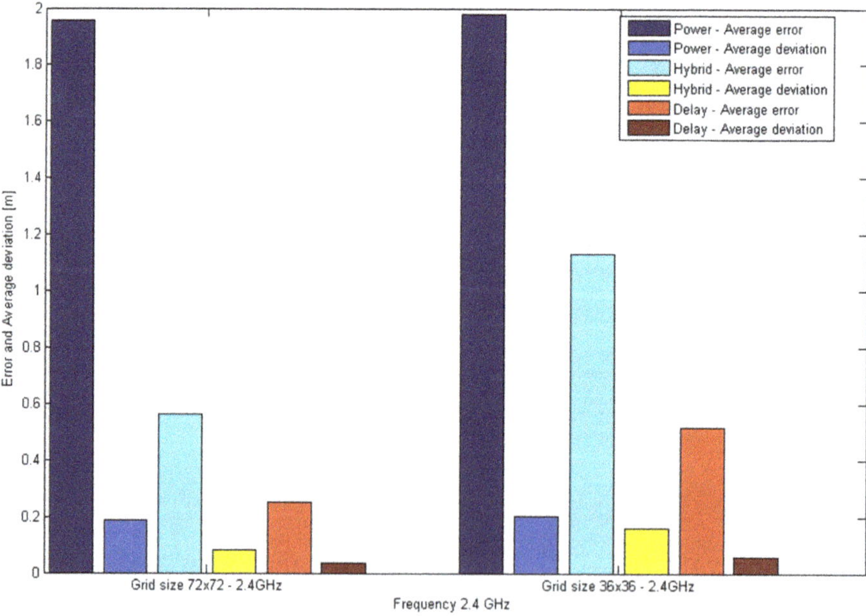

Figure 10. Detection methods comparison: Power, delay and hybrid detection at a frequency of 2.4 GHz.

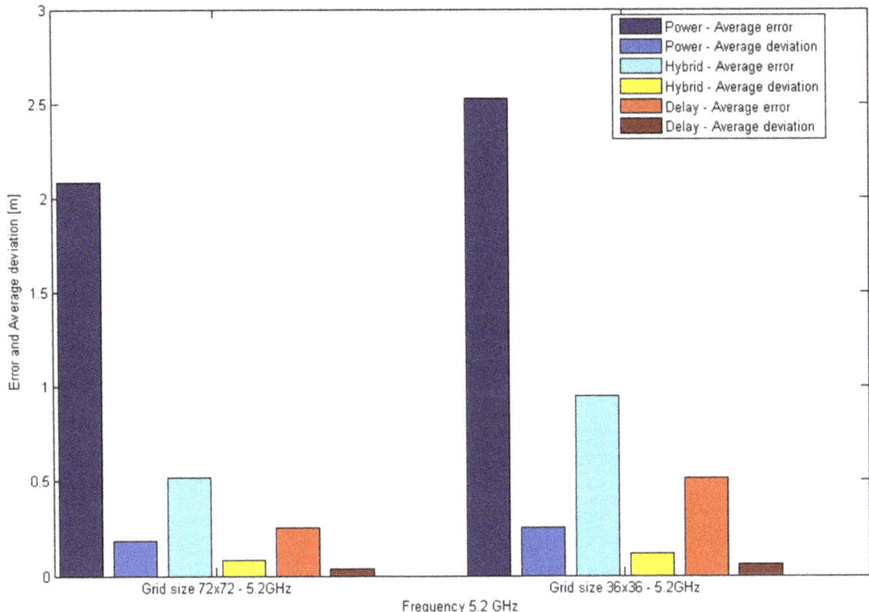

Figure 11. Detection methods comparison: Power, delay and hybrid detection at a frequency of 5.2 GHz.

From Figure 10, we observe that when the granularity of the fingerprints used in the off-line phase (radio-map) increases, the error committed is reduced. This reduction is independent of the detection method used as a cost function in the localization algorithm, but this reduction of the error is more noticeable when the method used is by ray-delays or hybridization. On the other hand, the reduction obtained by using power levels is not significant.

Figure 11 shows the same conclusions, but in this case, we also observe that the average error increases with the frequency. This is because the frequency is more than double (2.4 GHz versus 5.2 GHz) when comparing the Figure 11 with the Figure 10, or what is the same, the wavelength is less than half (0.057 m versus 0.125 m), what it means that the average error for the same grid resolution is increased (the power average error for a grid size of 36×36 m at 2.4 GHz was 1.98, similar to the 2.08 power average error for a grid size of 72×72 m at 5.2 GHz). In addition, the loss due to attenuation of the radio signal increases with the square of the power, which implies having a quarter of the power received with respect to the lower frequency band and contributing to the increase of the mentioned average error.

Table 3 shows information about the error and deviation values.

As it can be seen, the detection method based on power levels presents more error in the localization than using hybrid or delay detection methods. The pure delay detection method reaches the best results, because hybrid detection uses power levels information in the cost function.

The impact of frequency is not very relevant. For the same granularity, it has no impact on the delay detection method but improves the accuracy when it increases for power and hybrid methods.

The last two columns reflect the average error rate, which is improved by comparing the hybrid detection versus power detection and detection by ray-delays against power detection.

According to these results, the hybrid detection obtains an improvement in the average error of 71.28% for a mesh of 72×72 m at 2.4 GHz, 75% for a mesh of 72×72 m at 5.2 GHz, 32.82% for a mesh of 36×36 m at 2.4 GHz and 62.69% for a mesh of 36×36 m at 5.2 GHz. The detection method by ray-delays obtains an improvement in the average error of 87.17% for a mesh of 72×72 m at 2.4 GHz,

87.98% for a mesh of 72×72 m at 5.2 GHz, 74.24% for a mesh of 36×36 m at 2.4 GHz and 75.48% for a mesh of 36×36 m at 5.2 GHz.

Table 3. Indoor simulations results: Localization error when comparing the detection methods.

Grid size Frequency	Power	Hybrid	Delay	Hybrid vs. Power	Delay vs. Power
	Av_{Error}	Typ_{Dev}		Av_{Error} (%)	
72×72	1.9554	0.5621	0.2504	71.28	87.17
2.4 GHz	0.1871	0.0821	0.0366		
72×72	2.0844	0.5207	0.2504	75	87.98
5.2 GHz	0.1843	0.0841	0.0366		
36×36	1.9801	1.1337	0.5155	32.82	74.24
2.4 GHz	0.2029	0.1610	0.0572		
36×36	2.5275	0.9497	0.5155	62.69	75.48
5.2 GHz	0.2548	0.1142	0.0572		

5.2. Outdoor Scenarios

The analysed outdoor geometry corresponds to a real section of parking area at the Adolfo-Suarez Madrid-Barajas International Airport (Figure 12).

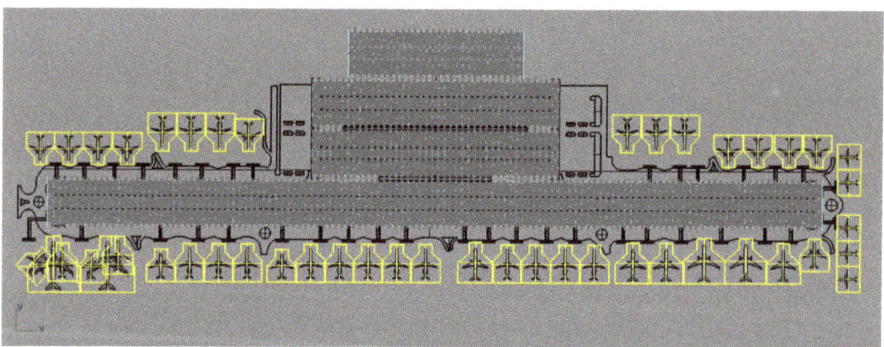

Figure 12. Outdoor scenario: Terminal 4 at Adolfo-Suarez Madrid-Barajas International Airport.

As part of the airport infrastructure, we propose enhancing the airport surveillance by providing several wireless access points in each boarding finger, which would be required to obtain the maximum coverage.

Three different scenarios were analysed. In all cases, to estimate the accuracy of the algorithm, 100 target mobiles were randomly located:

- Case 1: Grid size of 12×145 m

In the first scenario, the simulation area was 12×145 m for a grid resolution (distance between fingerprints) of 1 m. In this area, 1740 fingerprints and 5 APs at 2.4 GHz were employed (Figure 13).

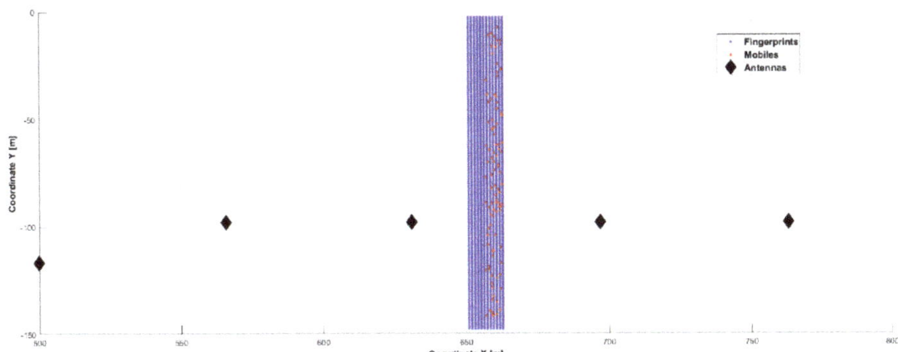

Figure 13. 2D view of Terminal 4 at Adolfo-Suarez Madrid-Barajas International Airport simulation area: A section of 12 × 145 m containing 1740 fingerprints, 99 mobiles to locate and 5 APs.

- Case 2: Grid size of 80 × 80 m

In our second scenario, the simulation area was 80 × 80 m for a grid resolution (distance between fingerprints) of 1 m. In this area, 6400 fingerprints and 5 APs at 2.4 GHz were utilized (Figure 14).

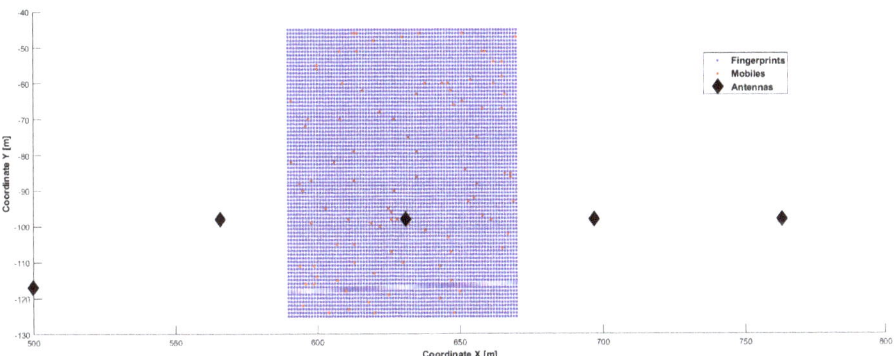

Figure 14. 2D view of Terminal 4 at Adolfo-Suarez Madrid-Barajas International Airport simulation area: A section of 80 × 80 m containing 6400 fingerprints, 99 mobiles to locate and 5 APs.

- Case 3: Grid size of 200 × 120 m

In this final scenario, the simulation area was 200 × 120 me for a grid resolution (distance between fingerprints) of 1 m. In this area, 24,000 fingerprints and 5 APs at 2.4 GHz were employed (Figure 15).

The results obtained from the outdoor simulations scenarios that were previously described are shown. In this case, four distance metrics (Euclidean, Manhattan, Bray-Curtis and Mahalanobis) and two fingerprinting detection methods (power levels and relative ray-delay between rays) were compared. Figures 16–18 show the simulation results for all cases.

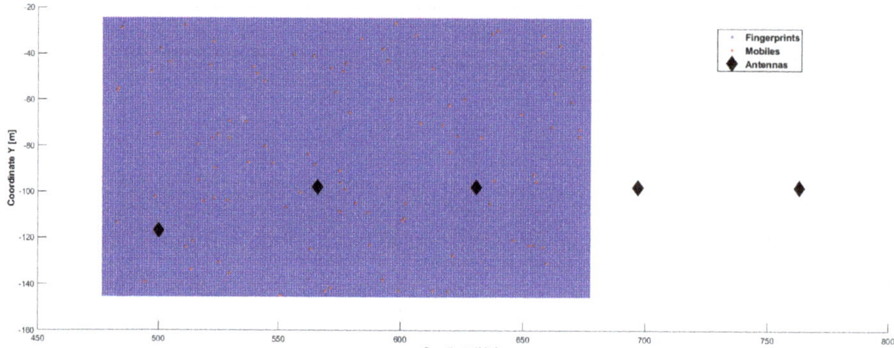

Figure 15. 2D view of Terminal 4 at Adolfo-Suarez Madrid-Barajas International Airport simulation area: A section of 200×120 m containing 24,000 fingerprints, 99 mobiles to locate and 5 APs.

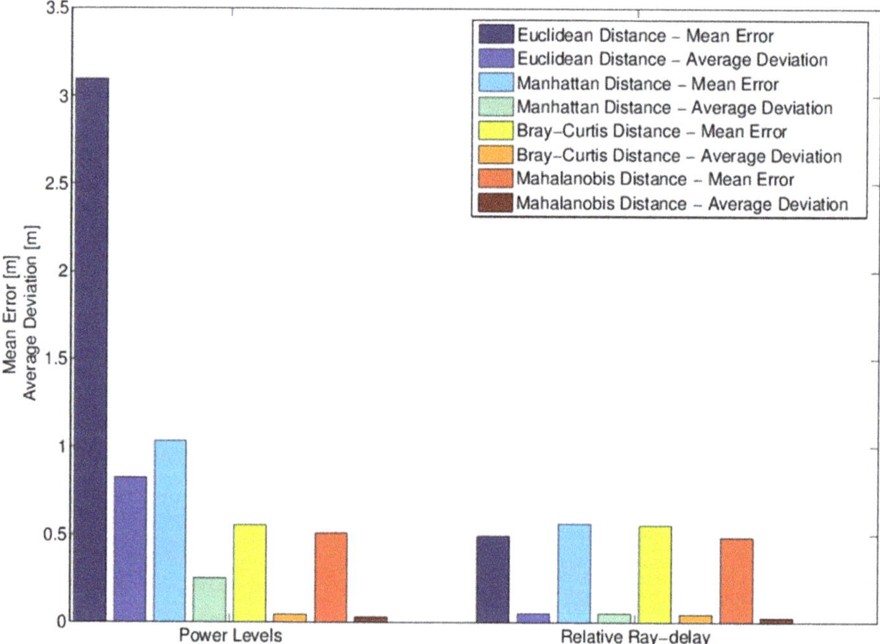

Figure 16. Case 1 results: Localization error when comparing power vs. delay detection methods with four distance metrics in a section of 12×145 m.

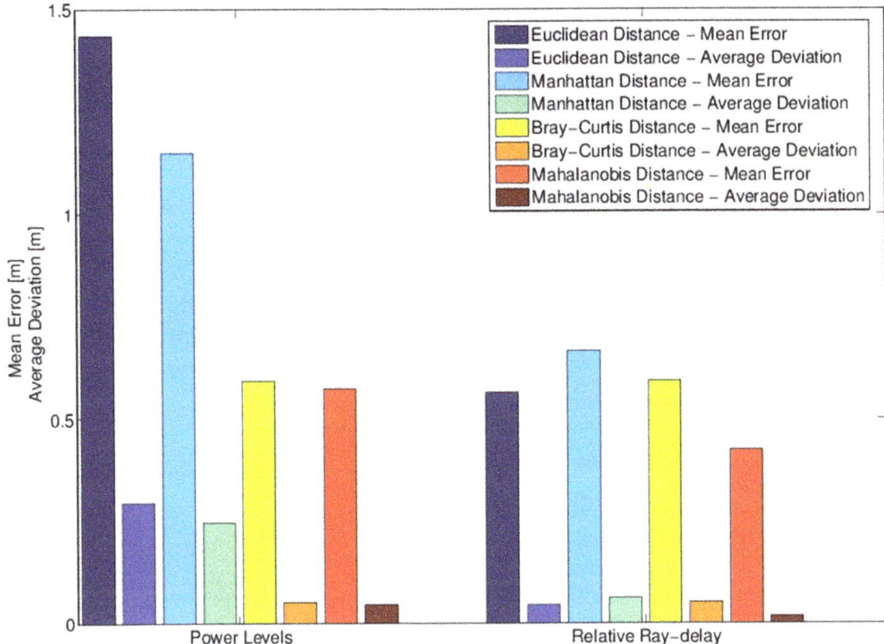

Figure 17. Case 2 results: Localization error when comparing power vs. delay detection methods with four distance metrics in a section of 80×80 m.

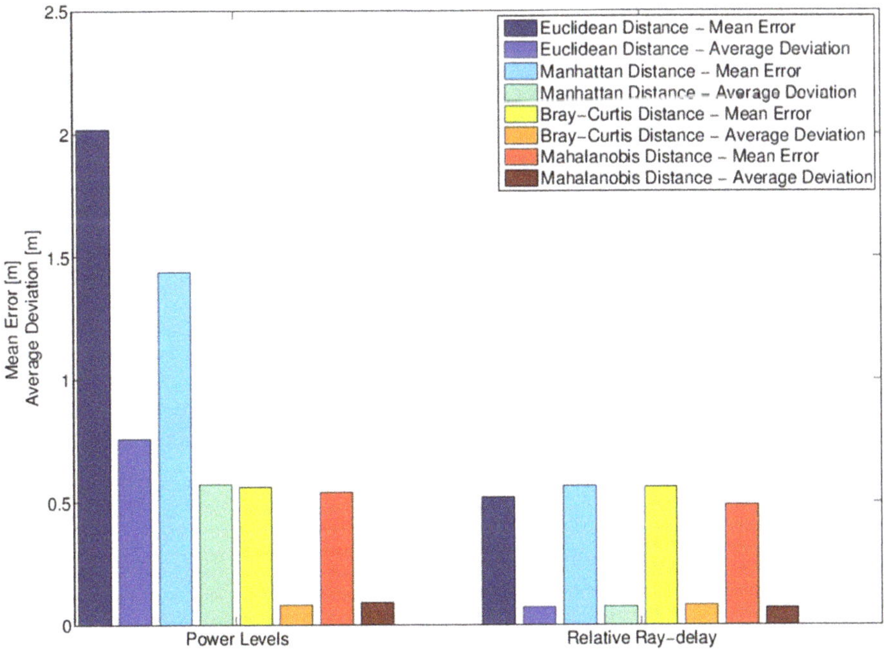

Figure 18. Case 3 results: Localization error when comparing power vs. delay detection methods with four distance metrics in a section of 200×120 m.

From Figure 16, we observe that if the relative delay between rays is used as a detection method, the average error decreases when comparing with the power level detection method. Moreover, if Mahalanobis distance is selected as a metric, we obtain the best results in terms of localization error reduction. Figure 17 shows similar conclusions, but in this case, the average error decreases as the granularity of the radio map increases. From Figure 18, we can obtain similar conclusions. Table 4 shows detailed information about the mean error values obtained in the simulations.

Table 4. Outdoor simulations results: Localization error when comparing power vs delay detection methods and distance metrics in each scenario.

Distance Metric	Grid Size					
	12 × 145 (m)		80 × 80 (m)		200 × 120 (m)	
	Detection Method					
	Power	Rays	Power	Rays	Power	Rays
	Mean Error (m)					
Euclidean	3.10	0.49	1.43	0.56	2.02	0.52
Manhattan	1.03	0.56	1.14	0.66	1.44	0.56
Bray-Curtis	0.55	0.55	0.59	0.59	0.56	0.56
Mahalanobis	0.50	**0.48**	0.57	**0.42**	0.80	**0.49**

In Table 4, we can analyse the numeric values of mean error obtained after simulating each scenario. According to Figures 16–18, the first two columns correspond with the first scenario (grid size of 12 × 145 m), the next two columns with the second scenario (80 × 80 m), and the last two columns correspond with the third scenario (200 × 120 m). In each scenario, we represent the mean error obtained using two different detection methods in the cost function of the localization algorithm, power and ray-delays. We observe how the relative delay between rays used as a detection method provides lower mean error than using power levels.

As we confirmed for indoor scenarios, from outdoor scenarios we observe that the relative ray-delay detection method also provides the best localization results for all analysed cases, while the size of the grid is not decisive. Mahalanobis distance metric improves the accuracy, compared with Euclidean, Manhattan or Bray-Curtis, because the relations between all samples used in the scenario are represented in its covariance matrix.

It can also be observed that the location errors obtained in the two scenarios, indoor and outdoor, are similar. In both cases there are multiple obstacles, so much information is obtained about ray-tracing and therefore the TDR factor in each fingerprint is distinctive for each of them.

6. Discussion, Assumptions and Further Work

Using a ray-tracing dataset, the main multipath effects are a result of the built environment and other static objects, all of which are captured in the modelling. The dynamic effects from mobile scatterers can be addressed by identifying these components via identifying the Doppler shift and eliminating them when comparing against the ray-tracing data. This issue will be the subject of further research in this project.

A description of the impact of electromagnetic noise on our simulations at the results level is not available. In an airport, for instance, many users who operate phones, other purposes of WIFIs, and other uncontrolled signals share the same spectrum (such as Bluetooth). To expect our system to have excellent performance even in these contexts, a more detailed noised scenario will be modelled in future research.

Reflective objects (aircrafts) and a more complex environment were considered in the analysis. However, a more detailed environment, which refers to the density of the multipath, will be further implemented to obtain more multipath components with increased path delay.

The scheme is not dependent on the delay values but a relationship between delay and other parameters, such as the BS location at which these parameters are recorded and the frequency, were analysed. Thus, the purpose of training (radio map) is to establish a relationship among the supporting vectors.

Referring to mobile scatterers in the environment, the sensitivity of the approach with mobile scatterers is not currently known and requires further study; thus, this issue would be the subject of future research. Possibilities exist to mitigate the effects of mobile scatterers by identifying these components via identifying the Doppler shift and eliminating when comparing against the ray-tracing data. An analogy exists with radar processing in removing clutter in this case. Thus, further processing using similar techniques to remove multipath components as a result of moving scatterers will be the subject of further research. In our study, ultra-wide band (UWB) systems that can resolve individual multipath effects are assumed.

In the same manner, a similar assumption for AOA estimation with multi-antenna systems was supplied. Massive MIMO systems have a large number of APs that are assumed to be capable of super-resolution AOA estimation. Both these assumptions are related to current and emerging 5G technologies and will be a part of further studies.

Time delay estimators, including bandwidth and SNR, will be required. Super-res algorithms can be used to achieve timing resolutions below the reciprocal of the bandwidth.

The method of localization proposed in this study is being evaluated against traditional techniques (TOA and TDOA) with the same framework and data. Thus, any limitation in the accuracy of the ray-tracing impacts are relatively similar since they employ the same test data in terms of the time delay measurements. Note that only the first arriving paths are applied in the methodology.

Not modelling the error due to bandwidth limitations may be the correct approach, which enables us to isolate the errors due to the estimators and directly compare their performance. Including additional errors due to the assumed bandwidth limitation may mask the difference in performance or make the results bandwidth-specific.

In a second phase, we intend to make simulations with random mobile objects. In any case, small objects and people will present reflections of very low power, so we expect that main reflections will be due to large static objects.

Finally, because validation was only done on a simulation approach, no experiments have been performed. Real experiments will be obtained through a pre-processing step supported by DSPs (Digital Signal Processor) to model the channel through an IFFT (Inverse Fast Fourier Transform) to obtain the impulse response and the multipath effects from the power spectrum [68], and a post-processing phase where to implement our algorithms after collecting data from real APs nodes. In order to obtain the relative delays with enough resolution in the time domain, these APs must be Ultra-Wide Band (UWB) devices because of their great bandwidth [69,70].

7. Conclusions

In this study, alternative detection methods that can be employed in the fingerprinting technique for a mobile location have been presented. These methods facilitate the analysis and design of indoor/outdoor location services. Using the NewFasant simulation tool, we obtain indoor/outdoor coverage by a complete 3D model combined with powerful ray-tracing techniques. This approach is very useful for avoiding complex and expensive measure campaigns and performing the off-line phase of the fingerprinting technique in a quick and accurate manner.

The location algorithm computes the Euclidean distance and other proposed metrics between the power levels and ray-delay samples available in unknown positions of a mobile and each fingerprint stored in the database. This study applies the multipath effects due to ray-tracing as a competitive advantage. Thus, a comparative study has been conducted between three detection methods that utilize different information as a cost function: power levels, relative ray delays and a hybrid method.

All algorithms were validated in indoor and outdoor scenarios using different granularities (grid size) and frequencies.

As a result of this study, we can conclude that the relative ray-delay detection technique, which presents better results in the indoor/outdoor location process than the power detection technique, is extensively applied in traditional Wi-Fi systems (based on the 802.11 standard). When information about certain fingerprints introduces ambiguities or errors in the localization using the powers levels, the hybrid detection method appears to be a suitable alternative. Due to the different effects caused by the ray-tracing, the proposed method uses the information about the relative ray-delay of the signals available in the fingerprints.

With respect to the metric, we conclude that conventional distance metrics, such as Euclidean or Manhattan, may not obtain the best location accuracy. The Mahalanobis distance metric improves the location accuracy when the geometry has irregularities that can been modelled among measures by the covariance matrix.

We can affirm that frequency does not have a significant impact on the accuracy of the results and that the grid size improves the results. Thus, the off-line phase of the fingerprinting technique can be carefully calibrated to obtain an optimized radio-map.

Author Contributions: A.D.C.-V. obtained the training data from the simulation tool and used MATLAB for the final implementation and validation. J.M.G.-P. was involved in the conceptualization and the methodology of the experiments. O.G.-B. have contributed to the electromagnetic theory and supervision of the simulations. J.L.C.-S. has contributed to the formal analysis and the bibliography references. All authors were involved in preparing the manuscript.

Funding: This research received no external funding.

Conflicts of Interest: The authors declare no conflict of interest.

References

1. Aernouts, M.; Berkvens, R.; Van Vlaenderen, K.; Weyn, M. Sigfox and LoRaWAN datasets for fingerprint localization in large urban and rural areas. *Data* **2018**, *3*, 13. [CrossRef]
2. Guo, X.; Li, L.; Xu, F.; Ansari, N. Expectation maximization indoor localization utilizing supporting set for internet of things. *IEEE Internet Things J.* **2019**, *6*, 2573–2582. [CrossRef]
3. Ciftler, B.; Kadri, A.; Guevenc, I. IoT localization for bistatic passive UHF RFID systems with 3-D radiation pattern. *IEEE Internet Things J.* **2017**, *4*, 905–916. [CrossRef]
4. Ko, H.; Ko, I.; Lee, D. Multi-criteria matrix localization and integration for personalized collaborative filtering in IoT environments. *Multimed. Tools Appl.* **2018**, *77*, 4697–4730. [CrossRef]
5. Hu, E.; Deng, Z.; Hu, M.; Yin, L.; Liu, W. Cooperative indoor positioning with factor graph based on FIM for wireless sensor network. *Future Gener. Comput. Syst.* **2018**, *89*, 126–136. [CrossRef]
6. Kumar, S.; Chaurasiya, V. A multisensor data fusion strategy for path selection in internet-of-things oriented wireless sensor network (WSN). *Concurr. Comput. Pract. Exp.* **2018**, *30*, e4477. [CrossRef]
7. Shit, R.C.; Sharma, S.; Puthal, D.; Zomaya, A.Y. Location of things (LoT): A review and taxonomy of sensors localization in IoT infrastructure. *IEEE Commun. Surv. Tutor.* **2018**, *20*, 2028–2061. [CrossRef]
8. Sotenga, P.Z.; Djouani, K.; Kurien, A.M.; Mwila, M.M. Indoor localization of wireless sensor nodes towards internet of things. *Procedia Comput. Sci.* **2017**, *109*, 92–99. [CrossRef]
9. Wang, J.; Bolic, M. Indoor localization using augmented ultra-high-frequency radio frequency identification system for Internet of Things. *Int. J. Distrib. Sens. Netw.* **2017**, *13*, 1550147717739814. [CrossRef]
10. Pedro, F.; Villa, K.; Elena, S. Wireless positioning in IoT: A look at current and future trends. *Sensor* **2018**, *18*, 2470.
11. Molina, B.; Olivares, E.; Palau, C.E.; Esteve, M. A multimodal fingerprint-based indoor positioning system for airports. *IEEE Access* **2018**, *6*, 10092–10106. [CrossRef]
12. Chen, L.; Thombre, S.; Järvinen, K.; Lohan, E.S.; Alén-Savikko, A.; Leppäkoski, H.; Bhuiyan, M.Z.; Bu-Pasha, S.; Ferrara, G.N.; Honkala, S.; et al. Robustness, security and privacy in location-based services for future IoT: A survey. *IEEE Access* **2017**, *5*, 8956–8977. [CrossRef]

13.	Sadowski, S.; Spachos, P. RSSI-based indoor localization with the internet of things. *IEEE Access* **2018**, *6*, 30149–30161. [CrossRef]

14.	Peng, L.; Dhaini, A.; Ho, P. Toward integrated cloud-fog networks for efficient IoT provisioning: Key challenges and solutions. *Future Gener. Comput. Syst.* **2018**, *88*, 606–613. [CrossRef]

15.	Saurabh, K.; Mukesh, A. Even localization in the internet of things environment. In Proceedings of the 7th International Conference on Advances in Computing and Communications, Cochin, India, 22–24 August 2017.

16.	Park, C.; Chang, J. Robust TOA source localization algorithm using prior location. *IET* **2019**, *13*, 384–391.

17.	Chen, H.; Ansari, N. Improved robust TOA-based localization via NLOS balancing parameter estimation. *IEEE Trans. Veh. Technol.* **2019**, *68*, 6177–6181. [CrossRef]

18.	Bellili, F.; Ben, A.; Affes, S.; Ghrayeb, A. Maximum likelihood joint angle and delay estimation from multipath and multicarrier transmissions with application to indoor localization over IEEE 802.11ac radio. *IEEE Trans. Mob. Comput.* **2019**, *18*, 1116–1132. [CrossRef]

19.	Alebrahim, M.; Zarka, N.; Orabi, M.; Dlaikan, S. GSM Navigation Techniques. 2015. Available online: https://www.researchgate.net/publication/288180429_GSM_Navigation_Techniques (accessed on 30 July 2019).

20.	Zhang, Y.; Zhang, L.; Han, J.; Ban, Z.; Yang, Y. An indoor gas leakage source localization algorithm using distributed maximum likelihood estimation in sensor networks. *J. Ambient Intell. Humaniz. Comput.* **2017**, *10*, 1703–1712.

21.	Sahlli, E.; Ismail, M.; Nordin, R.; Abdulah, N. Beamforming techniques for massive MIMO systems in 5G: Overview, classification, and trends for future research. *Front. Inf. Technol. Electron. Eng.* **2017**, *18*, 753–772.

22.	Gong, W.; Liu, J. RoArray: Towards more robust indoor localization using sparse recovery with commodity WIFI. *IEEE Trans. Mob. Comput.* **2018**, *18*, 1380–1392. [CrossRef]

23.	Gomez, J.; Cejudo, S.; Gonzalez, I.; Catedra, F. Application of high frequencies techniques for location systems. In Proceedings of the International Symposium on Electromagnetic Theory, Ottawa, ON, Canada, 26–28 July 2007.

24.	Huang, Y.; Hsu, Y. Indoor localization with adaptive channel model estimation in WiFi networks. In Proceedings of the International Conference on Mobile and Wireless Technology, Hong Kong, China, 25–27 June 2019.

25.	Krishnamurthy, P. WiFi Location Fingerprinting. In *Advanced Location-Based Technologies and Services*; Karimi, H.A., Ed.; CRC Press, University of Pittsburgh: Pittsburgh, PA, USA, 2017.

26.	Oguejiofor, O. Outdoor localization system using RSSI measurement of wireless sensor network. *Int. J. Innov. Technol. Explor. Eng.* **2013**, *2*, 1 6.

27.	Huang, H.; Lin, B.; Feng, L.; Lv, H. Hybrid indoor localization scheme with image sensor-based visible light positioning and pedestrian dead reckoning. *Appl. Opt.* **2019**, *58*, 3214. [CrossRef] [PubMed]

28.	Zhao, Y.; Zhou, H.; Li, M.; Kong, R. Implementation of indoor positioning system based on location fingerprinting in wireless networks. In Proceedings of the 4th International Conference on Wireless Communications, Networking and Mobile Computing, Dalian, China, 12–14 October 2008.

29.	Kaemarungsi, K.; Krishnamurthy, P. Properties of indoor received signal strength for wlan location fingerprinting. In Proceedings of the First Annual International Conference on Mobile and Ubiquitous Systems: Networking and Services (MOBIQUITOUS 2004), Boston, MA, USA, 26 August 2004.

30.	Lemelson, H.; Kjaergaard, M.; Hansen, R.; King, T. Error estimation for indoor 802.11 location fingerprinting. In Proceedings of the International Symposium on Location- and Context-Awareness, Tokyo, Japan, 7–8 May 2009.

31.	Martín, H.; Tarrío, P.; Bernardos, A.M.; Casar, J.R. Experimental Evaluation of Channel Modelling and Fingerprinting Localization Techniques for Sensor Networks. In *Advances in Soft Computing, Proceedings of the International Symposium on Distributed Computing and Artificial Intelligence, 22–24 October 2009*; Corchado, J.M., Rodríguez, S., Llinas, J., Molina, J.M., Eds.; Springer: Berlin/Heidelberg, Germany, 2009.

32.	Zhao, Y.; Zhou, H.; Li, M. A novel overlap area matching algorithm based on location fingerprinting in wireless networks. In Proceedings of the 5th International Conference on Wireless Communications, Networking and Mobile Computing, Beijing, China, 24–26 September 2009.

33.	Sertthin, C.; Fujii, T.; Ohtsuki, T.; Nakagawa, M. Multi-band received signal strength fingerprinting based indoor location system. *IEICE Trans. Commun.* **2010**, *93*, 1993–2003. [CrossRef]

34.	Arya, A.; Godlewski, P.; Melle, P. Performance analysis of outdoor localization systems based on RSS fingerprinting. In Proceedings of the 6th International Symposium on Wireless Communication Systems, Tuscany, Italy, 7–10 September 2009.

35. Kaemarungsi, K. Efficient design of indoor positioning systems based on location fingerprinting. In Proceedings of the International Conference on Wireless Networks, Communications and Mobile Computing, Maui, HI, USA, 13–16 June 2005.

36. Kaemarungsi, K.; Krishnamurthy, P. Modeling of indoor positioning systems based on location fingerprinting. In Proceedings of the IEEE INFOCOM 2004, Hong Kong, China, 7–11 March 2004.

37. Wu, Y.; Chen, P.; Gu, F.; Zheng, X.; Shang, J. HTrack: An efficient heading-aided map matching for indoor localization and tracking. *IEEE Sens. J.* **2019**, *19*, 3100–3110. [CrossRef]

38. Stella, M.; Russo, M.; Begusic, D. Location determination in indoor environment based on RSS fingerprinting and artificial neural network. In Proceedings of the 9th International Conference on Telecommunications, Zagreb, Croatia, 13–15 June 2007.

39. Farivar, R.; Wiczer, D.; Gutierrez, A.; Campbell, R. A statistical study on the impact of wireless signals' behavior on location estimation accuracy in 802.11 fingerprinting systems. In Proceedings of the IEEE International Symposium on Parallel & Distributed Processing, Rome, Italy, 23–29 May2009.

40. Hamza, L.; Nerguizian, C. Neural network and fingerprinting-based localization in dynamic channels. In Proceedings of the 6th IEEE International Symposium on Intelligent Signal Processing, Budapest, Hungary, 26–28 August 2009.

41. Lemelson, H.; Kopf, S.; King, T.; Effelsberg, W. Improvements for 802.11-based location fingerprinting systems. In Proceedings of the 33rd Annual IEEE International Computer Software and Applications Conference, Seattle, WA, USA, 20–24 July 2009.

42. Cherntanomwong, P.; Takada, J.; Tsuji, H.; Gray, D. New radio source localization using array antennas based on fingerprinting techniques in outdoor environment. In Proceedings of the Asia-Pacific Microwave Conference, Bangkok, Thailand, 11–14 December 2007.

43. Bshara, M.; Orguner, U.; Gustafsson, F.; Van Biesen, L. Fingerprinting localization in wireless networks based on received-signal-strength measurements: A case study on WIMAX networks. *IEEE Trans. Veh. Technol.* **2010**, *59*, 283–294. [CrossRef]

44. Bisio, I.; Lavagetto, F.; Marchese, M.; Sciarrone, A. GPS/HPS and Wi-Fi fingerprint-based location recognition for check-in applications over smartphones in cloud-based LBSs. *IEEE Trans. Multimed.* **2013**, *15*, 858–869. [CrossRef]

45. Azin, A.; Godlewski, P.; Melle, P. A hierarchical clustering technique for radio map compression in location fingerprinting systems. In Proceedings of the IEEE 71st Vehicular Technology Conference, Taipei, Taiwan, 16–19 May 2010.

46. Suroso, D.; Cherntanomwong, P.; Sooraksa, P.; Takada, J. Location fingerprint technique using fuzzy C-Means clustering algorithm for indoor localization. In Proceedings of the IEEE Region 10 Annual International Conference, Bali, Indonesia, 21–24 November 2011.

47. Kjaergaard, M.; Munk, C. Hyperbolic location fingerprinting: A calibration-free solution for handling differences in signal strength. In Proceedings of the 6th Annual IEEE International Conference on Pervasive Computing and Communications, Hong Kong, China, 17–21 March 2008.

48. Bolliger, P.; Partridge, K.; Chu, M.; Langheinrich, M. Improving location fingerprinting through motion detection and asynchronous interval labeling. In Proceedings of the International Symposium on Location and Context-Awareness, Tokyo, Japan, 7–8 May 2009.

49. Pirzad, N.; Yunus, M.; Subhan, F.; Abro, A.; Hassan, M. Location fingerprinting technique for WLAN device-free indoor localization system. *Wirel. Pers. Commun.* **2017**, *95*, 445–455. [CrossRef]

50. Soro, B.; Lee, C. A wavelet scattering feature extraction approach for deep neural network based indoor fingerprinting localization. *Sensors* **2019**, *19*, 1790. [CrossRef]

51. Steiner, C.; Wittneben, A. Low complexity location fingerprinting with generalized UWB energy detection receivers. *IEEE Trans. Signal. Process.* **2010**, *58*, 1756–1767. [CrossRef]

52. Kimoto, R.; Yamamoto, T.; Ishida, S.; Tagashira, S.; Fukuda, A. Evaluation of multi ZigLoc: Indoor ZigBee localization system using inter-channel characteristics. In Proceedings of the 11th International Conference on Mobile Computing and Ubiquitous Network, Auckland, New Zealand, 5–8 October 2018.

53. Wu, J.; Zhu, M.; Xiao, B.; Qiu, Y. The improved fingerprint-based indoor localization with RFID/PDR/MM technologies. In Proceedings of the IEEE 24th International Conference on Parallel and Distributed Systems (ICPADS), Singapore, 11–13 December 2018.

54. Li, Y.; Gao, Z.; He, Z.; Zhuang, Y.; Radi, A.; Chen, R.; El-Sheimy, N. Wireless fingerprinting uncertainty prediction based on machine learning. *Sensors* **2019**, *19*, 324. [CrossRef]

55. Zheng, H.; Gao, M.; Chen, Z.; Liu, X.; Feng, X. An adaptive sampling scheme via approximate volume sampling for fingerprint-based indoor localization. *IEEE Internet Things J.* **2019**, *6*, 2338–2353. [CrossRef]

56. Fang, S.; Lin, T. A dynamic system approach for radio location fingerprinting in wireless local area networks. *IEEE Trans. Commun.* **2010**, *58*, 1020–1025. [CrossRef]

57. Nicholaus, M.; Minga, D. A comparison review of indoor positioning techniques. *IJC* **2016**, *21*, 42–49.

58. Aksu, H.; Aksoy, D.; Korpeoglu, I. A study of localization metrics: Evaluation of position errors in wireless sensor networks. *Comput. Netw.* **2011**, *55*, 3562–3577. [CrossRef]

59. Gonzalez, I.; Lozano, L.; Cejudo, S.; Saez de Adana, F.; Catedra, F. New version of FASANT code. In Proceedings of the IEEE Antennas and Propagation Society International Symposium, San Diego, CA, USA, 5–11 July 2008.

60. Navarro, A.; Guevara, D.; Cardona, N.; López, J. Measurement-based ray-tracing models calibration in urban environments. In Proceedings of the 2012 IEEE International Symposium on Antennas and Propagation, Chicago, IL, USA, 8–14 July 2012.

61. Evangelo, M.; Andrew, N.; Geoffrey, S. Ray-tracing urban macrocell propagation statistics and comparison with WINNER II/+ measurements and models. In Proceedings of the Loughborough Antennas & Propagation Conference (LAPC), Loughborough, UK, 12–13 November 2012.

62. Junpe, M.; Tetsur, I.; Koshir, K. Ray-tracing system for predicting propagation characteristics on world wide web. In Proceedings of the 3rd European Conference on Antennas and Propagation (EuCAP 2009), Berlin, Germany, 23–27 March 2009.

63. Sáez de Adana, F.; Gutierrez, O.; Gonzalez, I.; Lozano, L.; Felipe Cátedra, M. *Practical Applications of Asymptotic Techniques in Electromagnetics*; Artech House Electromagnetics Series; Artech House: Boston, MA, USA, 2011.

64. Catedra, M.; Perez-Arriaga, J. *Cell Planning for Wireless Communications*; Artech House Publishers: Norwood, OH, USA, 1999.

65. Lozano, L.; Algar, M.; García, E.; González, I.; Catedra, F. Efficient combination of acceleration techniques applied to high frequency methods for solving radiation and scattering problems. *Comput. Phys. Commun.* **2017**, *221*, 28–41. [CrossRef]

66. Chen, Y.; Francisco, J.; Trappe, W.; Martin, R. A practical approach to landmark deployment for indoor localization. In Proceedings of the 3rd Annual IEEE Communications Society on Sensor and Ad Hoc Communications and Networks, Reston, VA, USA, 28 September 2006.

67. Perez-Ramirez, J.; Borah, D.; Voelz, D. Optimal 3-D Landmark Placement for Vehicle Localization Using Heterogeneous Sensors. *IEEE Trans. Veh. Technol.* **2013**, *62*, 2987–2999. [CrossRef]

68. Kent, H.; Michael, T. Comparison of predicted and measured multipath impulse responses. *IEEE Trans. Aerosp. Electron. Syst.* **2011**, *3*, 1696–1709.

69. Giuseppe, C.; Mai, T.; Luca, N.; Maria-Gabriella, B. Performance comparison of WiFi and UWB fingerprinting indoor positioning systems. *Technologies* **2018**, *6*, 14.

70. Sinan, G.; Zhi, T.; Georgios, G.; Hisashi, K.; Andreas, M.; Zafer, S. Localization via ultrawideband radios. *IEEE Signal Process. Mag.* **2005**, *22*, 70–84.

Article

Joint Balanced Routing and Energy Harvesting Strategy for Maximizing Network Lifetime in WSNs

Chih-Min Yu [1,*], Mohammad Tala't [2], Chun-Hao Chiu [3] and Chin-Yao Huang [3]

[1]   College of Artificial Intelligence, Yango University, Fujian 350015, China
[2]   Department of Electrical and Computer Engineering, National Chiao Tung University, Hsinchu 30010, Taiwan; moh.talat87.04g@g2.nctu.edu.tw
[3]   Institute of Electrical Engineering, Chung Hua University, Hsinchu 30012, Taiwan; howardchunhao@gmail.com (C.-H.C.); edcyh102@gmail.com (C.-Y.H.)
*   Correspondence: ycm@chu.edu.tw

Received: 22 May 2019; Accepted: 17 June 2019; Published: 18 June 2019

Abstract: Nowadays, wireless sensor networks (WSNs) are becoming increasingly popular due to the wide variety of applications. The network can be utilized to collect and transmit numerous types of messages to a data sink in a many-to-one fashion. The WSNs usually contain sensors with low communication ability and limited battery power, and the battery replacement is difficult in WSNs for large amount embedded nodes, which indicates a balanced routing strategy is essential to be developed for an extensive operation lifecycle. To realize the goal, the research challenges require not only to minimize the energy consumption in each node but also to balance the whole WSNs traffic load. In this article, a Shortest Path Tree with Energy Balance Routing strategy (SPT-EBR) based on a forward awareness factor is proposed. In SPT-EBR, Two methods are presented including the power consumption and the energy harvesting schemes to select the forwarding node according to the awareness factors of link weight. First, the packet forwarding rate factor is considered in the power consumption scheme to update the link weight for the sensors with higher power consumption and mitigate the traffic load of hotspot nodes to achieve the energy balance network. With the assistance of the power consumption scheme, hotspot nodes can be transferred from the irregular location to the same intra-layer from the sink. Based on this feature, the energy harvesting scheme combines both the packet forwarding rate and the power charging rate factors together to update the link weight with a new battery charging rate factor for hotspot nodes. Finally, simulation results validate that both power consumption and energy harvesting schemes in SPT-EBR achieve better energy balance performance and save more charging power than the conventional shortest path algorithm and thus improve the overall network lifecycle.

Keywords: Wireless Sensor Network; Dijkstra routing algorithm; load balance; power consumption; energy harvesting

---

1. Introduction

Wireless sensor networks (WSNs) have caused a lot of attention among scholars due to the growth of the Internet-of-Things (IoT) [1], and have achieved a wide success in daily life and industrial applications, such as smart grid [2], safety monitoring [3], and risk prevention [4]. The lifecycle of sensor nodes generally depends on the supplied battery power and numerous possible deployment methods to avoid unnecessary power depletion, including data processing, power control [5], energy efficient routing [6], etc. Long-term monitoring service is a key purpose for designing WSNs, but it is notably influenced by the network energy imbalance [7]. The challenge study issue of maximizing network lifecycle is preventing the energy imbalance, the condition happens when some nodes are exhausted, while others have plenty of residual energy, causing hotspot nodes in data transmission networks [8,9].

Directly sending packets with long distances from sensors to the data sink is unattractive, the routing protocols are usually designed in a multi-hop fashion, which can be classified into two cases: flat routing [10] and hierarchical routing [11]. In the flat routing case, routing paths from the sending sensors to sink are computed according to the minimum energy cost. The energy expense can be reduced by using shorter distance communication. In the hierarchical routing case, the network topology is constructed by clusters, and each cluster head (CH) is able to fuse packets with data correlations; hence, it is helpful for removing redundant packets, and decreases the total amount of packets, thus improves operation lifecycle.

Inventing a particular routing algorithm for energy efficient packet transmission is one main research direction. To consider the transmission costs and remaining battery power of sensors in [12], the authors formulate the network lifecycle maximum problem to a linear optimization problem and select the optimal shortest paths with the lowest cost, and near optimal lifecycle can be reached. To consider routing optimization in scenarios with obstacles in [13], a geographical routing is presented for the shortest path selection by using the Dijkstra algorithm. Each sensor is able to determine routing path according to location information of itself and the adjacent sensors, thus it is appropriate for the large scale WSNs. Another direction for balancing energy consumption is the modification on the typical flat and hierarchical routing schemes. To eliminate the unbalance power consumption of CH, the author in [14] studied how to determine the optimal cluster size for a decentralized hierarchical network. In addition, a General Self-Organized Tree-Based Energy-Balance (GSTEB) routing protocol is proposed in [15] to dynamically update the tree routing topology for the real-time situation and nodes can cooperate with each other via beacons to select the next node by their own routing decisions. Although all CHs produce an equal number of collected packets in [16], which leads the energy imbalance problem around the sink connection area (SCA), while causing underutilized energy for all the other sensors. Consequently, the research challenge issues in developing energy balance routing protocol are to equalize the power consumption among hotspots and achieve better power utilization of other sensors to improve the overall network lifecycle [17–19].

Another favorable technology to improve the lifecycle of WSNs is contributed by the recent energy harvesting models [20–23]. Each sensor captures the energy from the surrounding environment with energy harvesting abilities, including solar radiations or vibrations, and collects the energy into its own rechargeable battery. Therefore, it has a high probability to achieve the immortal WSNs when the energy harvest amount in each node is greater than the energy consumption amount for packet transmission. However, the major limitation of the method is the intermittent energy capture, which can make the network performance degraded.

To mitigate the irregular energy harvesting problem, wireless power transfer (WPT) transmits power distantly to the sensors, which provides an attractive opportunity for harvesting ambient energy [22]. Given the wireless charger, the energy transfer to the sensors can be controlled and make the network optimization in lifecycle extension. In [23], the authors optimize the network lifecycle performance under the condition that the energy consumption by each sensor is less than its harvested energy. Hence, the network operation lifecycle depends on not only how the wireless charger transfers the power to the sensors, but also how the sensors spend the reception power.

To jointly consider the energy balanced routing and energy harvesting benefits, a Shortest Path Tree with Energy Balance Routing strategy (SPT-EBR) is proposed in this article. In SPT-EBR, two methods are presented including the power consumption and the energy harvesting schemes to select the forwarding node according to the awareness factors of link weight. The SPT-EBR is a positive and early-intervention method, and includes the following contributions:

(1) To avoid excessive load concentration among SCA, a power consumption scheme is presented to generate the multiple shortest path trees in the early stage for data dissemination and distributes traffic load to the path trees according to the packet forwarding rate factor. The contribution is to

    improve the construction of routing solution and avoid the frequent routing update messages causing the excessive load concentration, thus prolonging the network lifecycle.

(2) In addition, the power consumption scheme is utilized to transfer some traffic load in the congested SCA to several sub-trees with the lowest load. This makes a part of paths with good load-balance level and high energy efficiency during the iteration process of the algorithm. As a result, each intra-layer from the sink can be expected to have the similar traffic load in terms of the balanced power consumption. The intra-layer is defined as the nodes with the same hop length to the sink.

(3) Using the late-remedy routing solutions [15,16], hotspot nodes are usually spatially distributed. With the assistance of the power consumption scheme, hotspot nodes can be transferred from the irregular location to the same intra-layer and this benefit makes the implementation of efficient wireless power charging over a large area is possible.

(4) To achieve the design aim of immortal WSNs, an energy harvesting scheme is proposed to jointly combine both the packet forwarding rate and the power charging rate factors together to make the energy charging efficient for the hotspots in the SCA. With the assistance of load balancing on SCA and energy harvesting, the contribution helps to accelerate the generation of the whole network balance routing solution, improving both the routing survivability and energy harvesting efficiency, as well as prolonging the network longevity.

The remainder of this paper is arranged as follows. The literature review is introduced in Section 2. Section 3 presents the problem statement and motivation. In Section 4, a network model is introduced and the SPT-EBR is presented including the power consumption and the energy harvesting schemes. In Section 5, the power consumption performances between the SPT-EBR and the Dijkstra algorithms are compared and the benefits of energy harvesting scheme are also demonstrated with computer simulations. Finally, conclusions are described in Section 6.

2. Related Work

2.1. Routing Based on Forwarding Factors

Many types of load-balanced transmission strategies have been proposed, including routing based on forwarding factors, hybrid power transmission, power control, etc. In the routing based on forwarding factors [16,24–27], routing paths are usually selected in accordance with several forwarding factors, such as the residual battery capacity, the distance to the sink, the degree of the nodes, and so forth.

The authors in [24] studied the ant colony optimization(ACO)-based path selection schemes by considering several factors, including the battery level of forwarding nodes, the distance to the sink, the travelled distance, and so on. Based on the one-dimensional queue network, the authors in [25] proposed an opportunistic routing protocol by considering forwarding factors with the distance to the sink and the remaining battery capacity of the sensors. Based on the multiple forwarding factors, a position-aware routing scheme [26] is proposed to consider the factors of the residual energy, node degree, and distance to the sink to balance the node energy and improve the network lifetime. A mixed transmission scheme is proposed in [27] that each node trades off between the multi-hop and single-hop transmission to extend the network lifetime by considering the forwarding factors including the link reliability, the degree of nodes and the remaining battery capacity In [16], two hybrid multi-hop/single-hop transmission strategies (power efficiency and power utilization strategies) with adaptive routing are proposed to extend the network lifetime of wireless sensor networks (WSNs). The power efficiency strategy aims to minimize the overall power consumption and the power utilization strategy endeavors to minimize the maximum power consumption among hotspots. In addition, the adaptive routing method is a late-remedy method to consider the channel status in each operation round.

2.2. Wireless Power Transfer

With recent energy harvesting models, optimizing data routing is a common approach to reduce energy consumptions of the nodes and improve the lifecycle of WSNs. Considering the power transfer task, the chargers construct power beams [28] to improve the wireless power transfer (WPT) efficiency, such that more power could be harvested by the sensors. To maximize the power reception by the sensor, the present approaches are based on the dominant channel eigenvector of energy beam [29,30]. Even the energy beamforming is optimized; if the nodes consume energy inefficiently, the power would be wasted and the network lifecycle performance will be degraded. To maximize the network lifecycle and achieve the immortality of the WSNs, the authors propose to reduce the energy consumption of sensors by optimizing the data routing among the sensors and to improve the WPT efficiency by optimizing the energy beamforming of the chargers [31].

Energy harvesting and WPT are capable of relieving the battery limitation of sensors, an amplify-and-forward relay network (AF-RN) is proposed in [32]. The strategy is a joint power control and energy transfer scheme to maximize the throughput by considering the energy causality constraints. In [33], an AF-RN with energy harvesting (EH) source and relay nodes is investigated. To consider the energy arrival profiles and the energy cooperation between EH nodes, a joint power control and transfer is designed to maximize the total data rate, subject to energy causality and battery storage constraints. With the total source transmission power budget and energy-causality constraints, the authors in [34] formulated an energy efficiency maximization (EEM) optimization problem for the multi-user multicarrier energy-constrained amplify-and-forward (AF) multi-relay network. Under the simultaneous wireless information and power transfer (SWIPT) model, each forwarding node is solely powered by the source nodes, employing an energy harvesting time-switching (EHTS) scheme to harvest the energy via the radio-frequency (RF) signal transmitted from the source nodes. In addition, a suboptimal and best relay selection algorithm is also examined to trade-off between complexity and performance.

The above energy balance approaches seem to be comprehensive; however, they are just passive and late-remedy solutions based on multiple forwarding factors to determine an appropriate routing path. On the other hand, the above efficient wireless charging schemes consider the efficient routing node selection instead of the balanced routing node determination. To jointly consider the design issues on the balanced routing and the effective WPT, a SPT-EBR is proposed to improve the load balance and thus improve the WPT efficiency. The SPT-EBR is an early-intervention method to make the load balancing in the SCA layer and all the other intra-layers. The proposed method can also avoid the frequent routing update messages and the excessive load concentration in the following maintenance phase. After load balance in each intra-layer, the energy harvesting scheme can be applied to efficiently charge for the hotspot nodes.

3. Problem Statement and Motivation

3.1. Problem Statement

When considering traditional multi-hop routing with many-to-one traffic patterns, unbalanced energy consumption is an inherent problem in WSNs. However, the multi-hop routing method can achieve better energy efficiency than the single-hop routing method in networks. In the multi-hop routing method, each sensor node can transmit and forward packets to the data sink through the shortest path, either occasionally or periodically. After a period of operation, hotspot nodes will be generated when the high traffic routing paths are converged and most of the traffic load congested at some specific nodes. Typically, sensors near the sink forward a larger amount of traffic load than sensors that are far from the sink. Therefore, the hotspot nodes near the data sink deplete the battery faster than other sensors. Since the network operation lifecycle is usually defined as the occasion when a battery of the first node is exhausted, the remaining energy capacity of other sensors can be regarded as underutilized energy, and the energy consumption discrepancy results in the considerable network lifetime degradation.

A traditional hierarchical WSN routing scenario that includes 81 clusters with one data sink and an evenly distributed two-dimensional network is shown in Figure 1. According to the energy capacity, a cluster head (CH) is selected from the sensors in each cluster. In the cluster, each CH is responsible to manage the sensors, merges all the sensing data from sensors, and sends the collected data to the sink. In order to achieve balance energy utilization, the node with the largest remaining battery capacity is designated as a CH in turn.

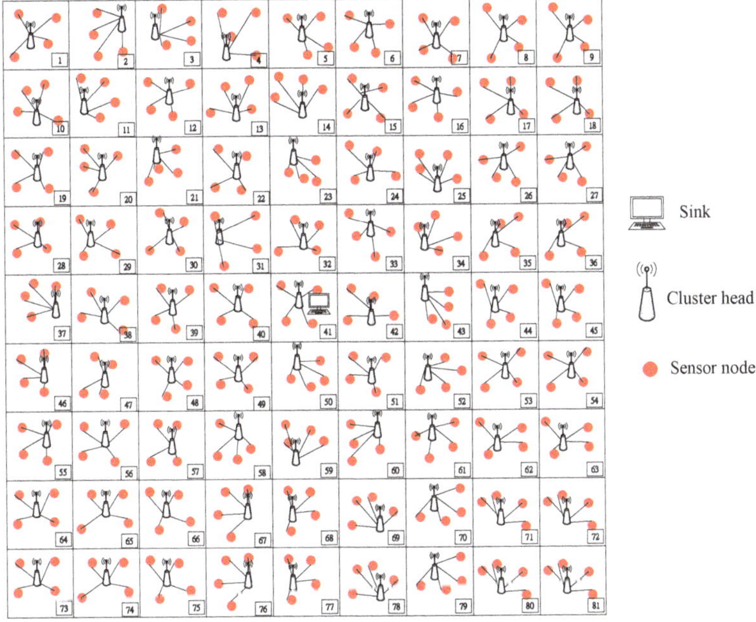

Figure 1. An 81-cluster example illustration for wireless sensor networks (WSN) deployment.

In the WSNs example, it is assumed that the nodes are deployed in a uniform distribution; the communication distance between two adjacent nodes is d. Each node transmits the sensing data to the sink through the forwarding node along the shortest path. The power consumption is assumed to be proportional to d^2 in the path loss model. Note that the shortest path from each node to the sink is calculated by the Dijkstra algorithm. From the perspective of data application, the traffic generated in each node is periodically sent to the target sink. After a specific network operation cycle, each node calculates the number of sending and forwarding packets, and then sends the statistics of the total transmitted packets along with data packets to the sink. Finally, the statistics of each node are calculated by the sink to estimate the number of paths forwarded.

In Figure 2, the average forwarding path load amount is illustrated. The 36 nodes in the red area are mainly forwarded through the node 32, the 20 nodes in the blue area are mainly forwarded by the node 40, the 20 nodes in the green area are mainly forwarded via the node 42, and the four nodes in the purple area are mainly forwarded by the node 50. Even though both nodes 32 and 50 are directly connected to the sink, the number of packets forwarding via node 32 is nine times that of node 50. Therefore, an interesting phenomenon is observed which leads to uneven traffic load among the sink in the network even with a uniform node distribution and each node has a uniform packet transmission rate. Specifically, the power consumption rate is unbalanced in the sink connectivity area (SCA), where the power consumption rate can be calculated by the total number of packets sent and forwarded in the period of each node.

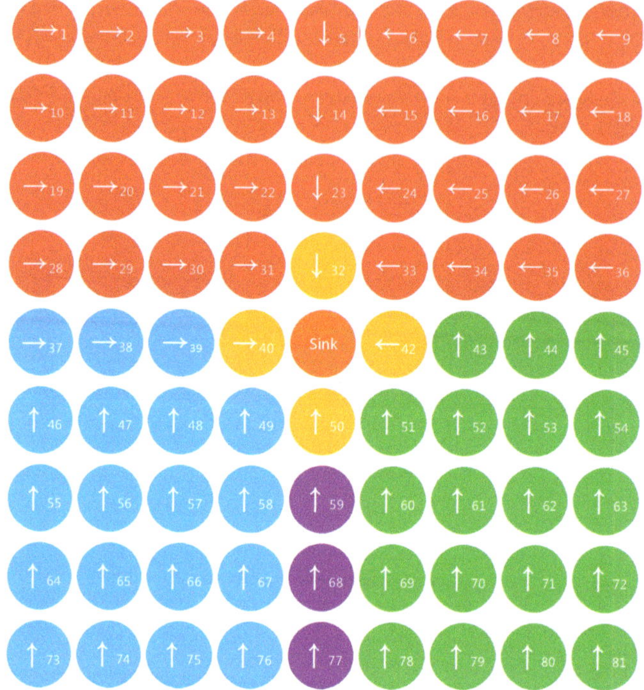

Figure 2. An illustrated example of packet forwarding amount for the sink connection area (SCA).

In the SCA, it is expected that the four primary forwarding nodes 32, 40, 42, and 50 achieve the approximate power consumption rate per operation cycle. Unfortunately, the Dijkstra algorithm uses the first shortest path and does not consider the following shortest paths; a non-uniform forwarding packet load distribution phenomenon was found in the SCA. In addition, the battery capacity of node 32 was first exhausted because the number of packets forwarded by the primary forwarding node adjacent to the data sink presents an uneven load distribution. The earliest exhausted node also causes the sub-area network to be unserved, thereby reducing the lifetime of the entire network. As a result, there are 35 nodes with good battery conditions, but these nodes cannot send data packets to the data sink through the primary forwarding node 32. As a result, the energy resource of the other nodes is not fully utilized in terms of the overall network underutilization. To improve the routing survivability, the highest traffic load can be transferred to several sub-trees with the least load, in order to speed up load balance among different nodes of the network.

3.2. Motivation

With the same link weight in each path, the Dijkstra algorithm computes the shortest path for all nodes of multi-hop network initially. After that, the routing path of each node changes as the path cost variation. To mitigate the power consumption of the hotspot nodes, the nodes with lighter traffic can share the forwarding packets if the hotspot nodes increase their path costs. At the same time, because the packet forwarding amount of hotspots next to the sink is very large, the packet forwarding rate is much higher than other nodes. Therefore, it is possible to design an energy balanced routing strategy by the amount of packet forwarding in each node and the packet forwarding rate can be utilized as a forward awareness factor of path cost or link weight. It can be expected that after several transmission cycles, the SCA and other nodes in each communication layer centered at the sink can achieve the effect of load balance in terms of the energy balance of each network layer.

In Figure 2, four main forwarding nodes directly connected to the sink can be observed that the number of forwarding packets via node 32 is nine times than that of node 50. We aimed to design a method with the packet forwarding rate for the path cost and to achieve the effect of load balancing of the four main forwarding nodes. Taking node 9 as an example in Figure 1, node 9 can reach sink 41 either from the primary node 32 or 42. Originally, the resulting shortest path is 9→8→7→6→5→14→23→32 by the Dijkstra algorithm. After a certain period of operation, if the forwarding amount of node 32 is greater than the other primary nodes in SCA, the path cost of node 32 is increased and an alternative path 9→8→7→15→25→34→33→42 could be generated. Thus, the packets originally forwarded by the node 32 will be forwarded by the node 42 afterward and the traffic load of node 32 can be mitigated. After that, when the forwarding amount of the node 42 is relatively increased and the path cost of the node 42 will be increased. As a result, the shortest path for node 9 can be alternative switched either via node 32 or node 42.

On the other hand, the forwarding traffic load via node 42 such as nodes 78 to 81 will also be transferred to node 50, since it has a small number of forwarding packets than node 42. After changing the path cost of hotspots to update the new routing paths of other sensors, our first aim is to design a uniform forwarding packet rate for the main forwarding nodes and all the other communication layers. In addition, the hotspot nodes location can be transferred from the irregular distributed area into the SCA intra-layer. Finally, the ideal balanced shortest tree of multi-hop networks can be seen in Figure 3. Based on the balanced shortest tree routing, the energy harvest capability is jointly deployed for each sensor to effectively utilize the harvested energy and towards to achieve the main goal for an immortal WSN.

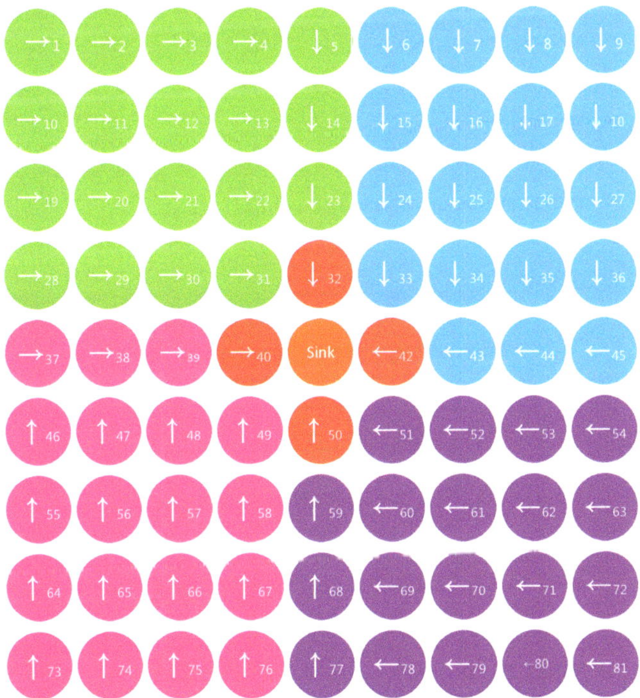

Figure 3. The ideal balanced shortest tree of multi-hop networks.

4. The Proposed Strategy

4.1. The Network Model

It is assumed that all sensors are evenly distributed in a two-dimensional region R with a radius of d for the network model. With unlimited energy, a static data sink is located at the center of the two-dimensional region. Each sensor is located within a cluster and transmits the sensed packet to its associated cluster head (CH). Then each CH aggregates and transmits packets along with the constructed shortest paths to the data sink. To evenly assign the CH among sensors, the CH is alternated with an equal probability in the same cluster and all sensors have the same amount of initial energy E_{init} [9].

In the network application, each sensor collects parameters such as humidity, temperature, air quality, or other related events at the outside environment and the fused data can be forwarded to the sink via the CHs along the transmission path. Where m is a variable that is represented the total amount of sending packets and forwarding packets from CH u to the sink. In each data cycle, p_k is the probability that a CH fuses the kth event and generates number of packets g_k. As a result, the packet generation rate r for any CH u in each data cycle can be expressed as:

$$r_u = \sum_{k=1}^{m} p_k g_k \tag{1}$$

To simplify the power computation model, the transmission power is the main factor to be taken into consideration in the path loss model since it spends a larger amount of power than the reception and idle situations. Hence, the power consumption for the multi-hop transmission can be represented as $P_M(r_u)$. Consequently, the network lifecycle of CH node u is expressed by:

$$N(u) = \frac{E_{init}}{P_M(r_u)} \tag{2}$$

where E_{init} is the initial amount of battery capacity of each sensor and the lifecycle is defined as the network operation round from sensors placement till the first sensor depletes its battery capacity [16].

4.2. The Proposed SPT-EBR Strategy

In order to prolong the lifecycle of WSNs with node energy efficiency and energy balance, this paper proposes two novel schemes by using forward awareness factors as path cost to select the next node for packet transmission. One is the power consumption scheme, and the other is the energy harvesting scheme. The power consumption scheme considers the packet forwarding rate as the forward awareness factor for each node to alter the desired routing path. The packet forwarding rate is proportional to the power consumption rate and it is considered to keep the sensors using the shortest path, thus making the packets transmission efficient. Simultaneously, the power consumption between the hotspots is also balanced, achieving the effect of both packet transmission efficiency and load balancing design goal.

In general, hotspot nodes are usually spatially distributed. With the assistance of the power consumption scheme, hotspot nodes can be transferred from the irregular location to the same intra-layer, and this benefit makes the implementation of efficient wireless power charging over a large area is possible. In the energy harvesting scheme, it is assumed that the network nodes have the capability to capture energy. The role of the charging station can be deployed by a nearby data sink or other RF base stations, and the sensing node can be charged in a random model or a constant model. The power charging rate in each node is combined with the packet forwarding rate as the forward awareness factor to calculate the path cost for next node selection, achieving the design goal of energy balance and permanent utilization WSNs.

4.2.1. The Power Consumption Scheme

In the power consumption rate scheme, time resource is allocated as a time slot that operates periodically to avoid frequent paths updating. In the beginning, the sink calculates the shortest path from all nodes to itself by the Dijkstra algorithm. The shortest path is then passed from sink to each sensor to establish a minimum spanning tree routing. Packets transmission with the shortest path will make the transmission of the entire network more energy efficient. In each time slot, each node calculates whether its own packet transmission rate is greater than the threshold value. The threshold value is determined by the sink and can be defined as the average of the overall packet forwarding rate. Then, the sink passes the threshold value to all nodes together when the routing paths are transmitted. After that, each node can update the path cost according to the threshold value.

In each time slot, if the total amount of packet transmission value P is greater than the threshold T in each node, the path cost is increased by one. When the packet transmission amount P of the node is less than or equal to T, the value of path cost keep the same. In order to avoid communication overhead, path cost can be transmitted to the sink along with the sensing data. When the sink receives the data, if the path cost changes, the sink recalculates and updates the path to the nodes with high traffic load.

In each time slot, the node with a faster packet transmission rate can mitigate forwarding packets by increases the path cost. Let other nodes near the sink share the network traffic load and achieve the traffic load balance in each layer of the network. Repeat the load balancing operation can balance the transmission packets of the hotspot nodes since the hotspots can share the forwarding packets in turn from all the other sensors. Distributing the traffic load allows network load balancing to be achieved while letting the global nodes to fully utilize their energy. Alternatively, the shortest path transmission also minimizes overall power consumption for efficient energy utilization. Generally, the power consumption scheme achieves both energy efficiency and energy balance to extend the overall network lifecycle. The detail design flowchart of the power consumption scheme is shown in Figure 4.

4.2.2. The Energy Harvesting Scheme

Wireless renewable sensor network is another major research issue in this paper. Assuming that the nodes have energy capture capabilities, the sensing nodes can replenish energy from the energy radiated by the solar or RF base stations. Here, an integration of power charging rate and data forwarding rate is proposed as shown in Figure 5. The operation calculation cycle is scheduled from *t(0)*, *t(1)*, *t(2)* to *t(n)* time slots. At each operation cycle, each node has the opportunity to capture random ambient energy *a(i)*, which represents the power capture rate of the *i*-th node. The packet transmission rate can be converted into the function of the power consumption rate in each node, and the packet transmission rate is proportional to the power consumption rate. Therefore, the packet transfer amount P is converted into a corresponding power consumption function *p(i)*, which represents the power consumption of the *i*-th node at each time slot. The initial battery capacity of each node is $b(0)$, and the new battery capacity is updated in each time slot. The current battery capacity $b(i)$ can be obtained from the previous battery capacity $b(i-1)$, plus the current charging energy $a(i)$ and minus the power consumption rate for packets transmission. The current battery capacity can be formulated in Equation (3):

$$b(i) = b(i-1) + a(i) - p(i)$$
$$i = 1, 2, 3 \ldots n$$

(3)

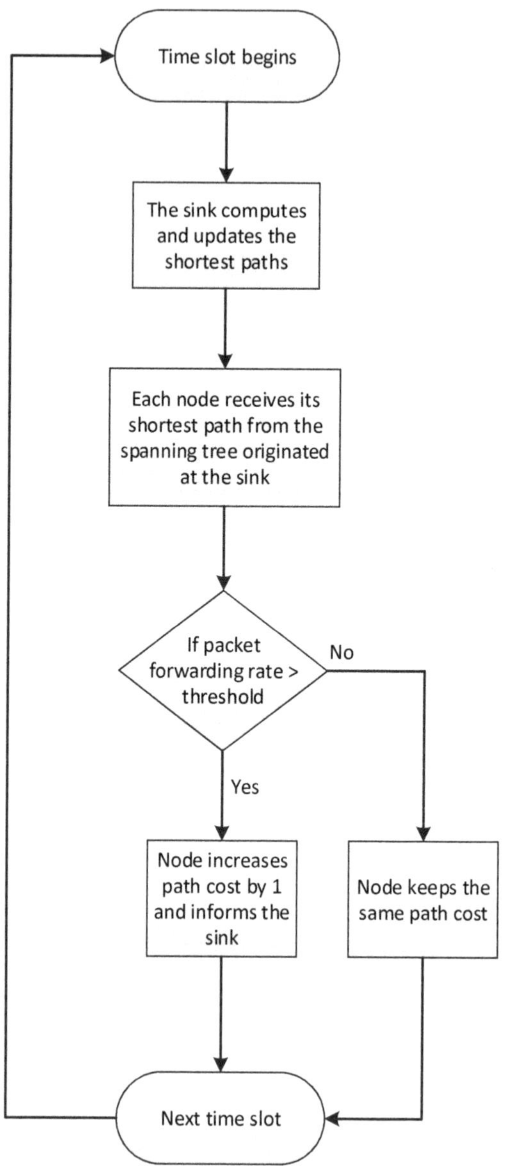

Figure 4. The illustrated flowchart of packet forwarding rate scheme.

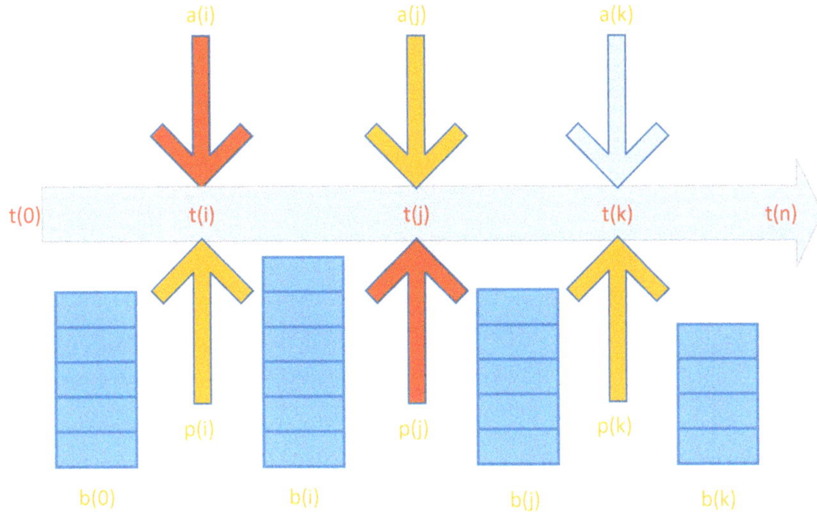

Figure 5. The illustration of the battery charging rate of the energy harvesting scheme.

From Equation (3), the charge rate of the battery capacity unit Δb can be obtained. The Δb represents the deviation of the current battery capacity with the battery capacity in the previous cycle $b(i-1)$. It is also the result of the rate of harvested energy minus the rate of consumed power in each node.

To achieve the immortal WSNs, an additional parameter called battery charging rate Δb is introduced to be the forward awareness factor. If the parameter Δb is greater than 0, it means that the battery of the sensing node is always in the positive charging state and will not be affected by the power consumption of packet transmission. The battery charging rate is the main considering parameter to update the path cost of each node. If the parameter is less than or equal to 0, it means that the power consumption rate of the node is faster than the power charging rate, and the power consumption rate is regarded as the main consideration factor. Therefore, the path cost design of the energy harvesting scheme can be considered together with both the packet forwarding rate and energy harvest rate.

The operation of the forward awareness factor as the path cost of power harvesting scheme can be described as follows. In each time slot, it is required to consider whether the Δb parameter is a positive value. If Δb is greater than 0, the path cost still keeps the same. In contrast, if the charging rate is less than or equal to 0, it is considered to follow the design principle of the first power consumption scheme. When the power consumption rate $p(i)$ in each node is greater than the average power consumption threshold value $p_{avg}(i)$, the path cost is increased by 2. If the $p(i)$ is less than or equal to the $p_{avg}(i)$, the path cost is incremented by 1. The detail design flowchart of the energy harvesting scheme are shown in Figure 6.

In real operation, the extra cost is generated to deploy the energy harvesting device and it will increase some subordinate costs. However, it can effectively extend the lifecycle of the sensing network and eliminate the requirement to replace the battery. To joint consider the energy balanced routing and energy harvesting benefits, the design aim of the immortal WSNs can be achieved when the energy harvesting rate in the hotspots is greater than energy consumption rate for packet transmission.

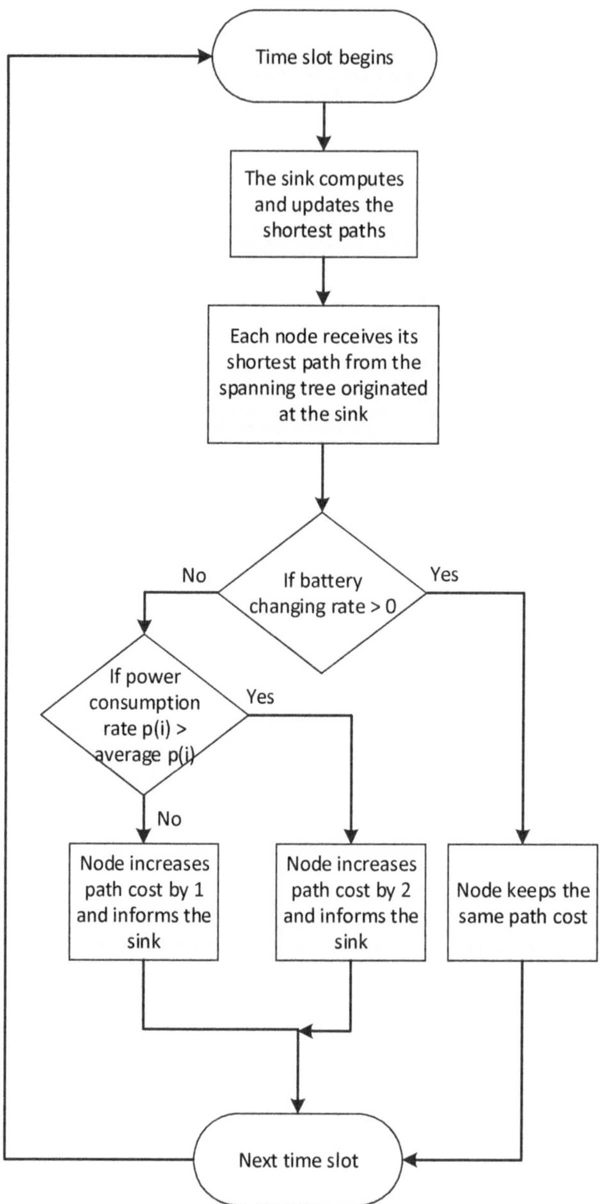

Figure 6. The illustrated flowchart of power harvest scheme.

5. Simulation Results

To validate the effectiveness of the proposed strategy, the network performances of energy routing protocols including the proposed SPT-LBR and the conventional Dijkstra algorithms are both investigated. To estimate the energy routing performances, a two-dimensional planar network with one data sink (Identification, ID 41) and 80 CHs nodes are studied here. The 80 CHs are even distributed in a given operation area and each cluster owns several sensors to detect the cluster area.

The four primary nodes in the SCA are 32, 41, 42, and 50. The SPT-EBR and the Dijkstra algorithms are both applied to compute the shortest paths with various forward awareness factors including the packet forwarding rate and the energy harvesting rate. With regard to the packet transmission, each sensor sends packets to the CHs and each CH fuses data as well as forwards them to the sink periodically. Two performances including the power consumption rate distribution and the network lifetime are compared between the power consumption scheme, power harvesting scheme, and the Dijkstra algorithm.

Figure 7 shows the power consumption rate distribution of the power consumption scheme in the red line and the Dijkstra algorithm in the blue line are both examined. In the Dijkstra algorithm, node 32 has the greediest power consumption compared to the other primary nodes, and is nine times than node 50 in average power consumption. In the power consumption scheme of SPT-EBR, the highest power consumption is at nodes 50, 1.4 times than node 41 in average power consumption. As a result, the power consumption scheme achieves better energy utilization with the load balance factor design. From the perspective on hotspot nodes distribution, the hotspot nodes 14, 23, 32, 40, and 42 are spatially distributed in the inter-layers by the Dijkstra routing. With the assistance of the power consumption scheme, the hotspot nodes can be evenly distributed around the sink connectivity area. In addition, the power consumption rate distribution in each intra-layer is more even than the conventional scheme.

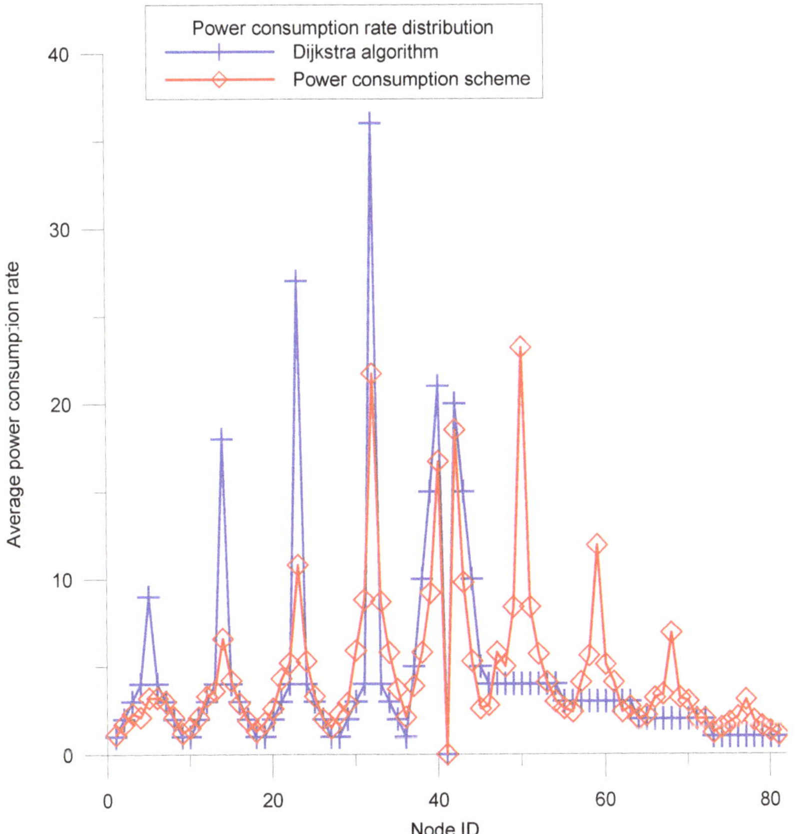

Figure 7. The power consumption rate distribution of the power consumption scheme in Shortest Path Tree with Energy Balance Routing strategy (SPT-EBR).

In Figure 8, the power consumption rate distribution of the larger network with 169 nodes for both the power consumption scheme and the Dijkstra algorithm are investigated. In the Dijkstra algorithm, node 72 encounters the highest power consumption rate than all the other primary nodes and there is 13 times than node 98 in average power consumption. This may lead to an inevitable energy unbalance problem in hotspot nodes as the network size grows. In the power consumption scheme of SPT-EBR, the highest power consumption rate is occurred at nodes 98 and there is 1.56 times than node 86 in average power consumption. With the load balance factor to select the next forwarding node, the power consumption scheme achieves better energy utilization than the Dijkstra algorithm. In addition, the energy unbalance problem can be controlled with the packet forwarding rate factor in the power consumption scheme.

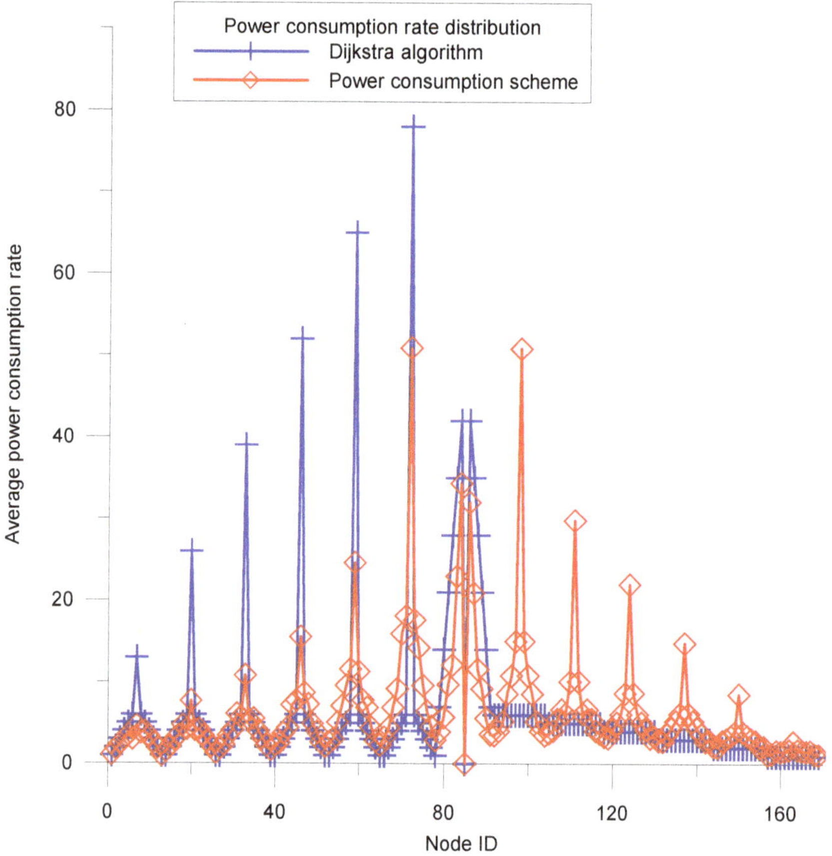

Figure 8. The power consumption scheme for larger SPT-EBR networks.

To evaluate the energy harvesting scheme, a uniform distribution of energy harvesting mode is introduced here to perform the simulation. The energy harvesting strategy includes the full nodes charging and the partial nodes charging strategies. The partial nodes charging strategy is designed to charge the 20% heavy traffic nodes and save the deployment cost in realizing the energy charging. The power charging value is uniformly distributed from 0 to 18 for the energy harvesting scheme and the average battery charging rate of the overall nodes is 7. Figure 9 shows the power consumption rate distribution of the energy harvesting scheme and both schemes achieve the identical power consumption for the four primary hotspot nodes. In general, the full nodes charging method achieves better performance than the partial nodes charging approach for most of the other nodes in power consumption since it spends more operation cost for energy charging.

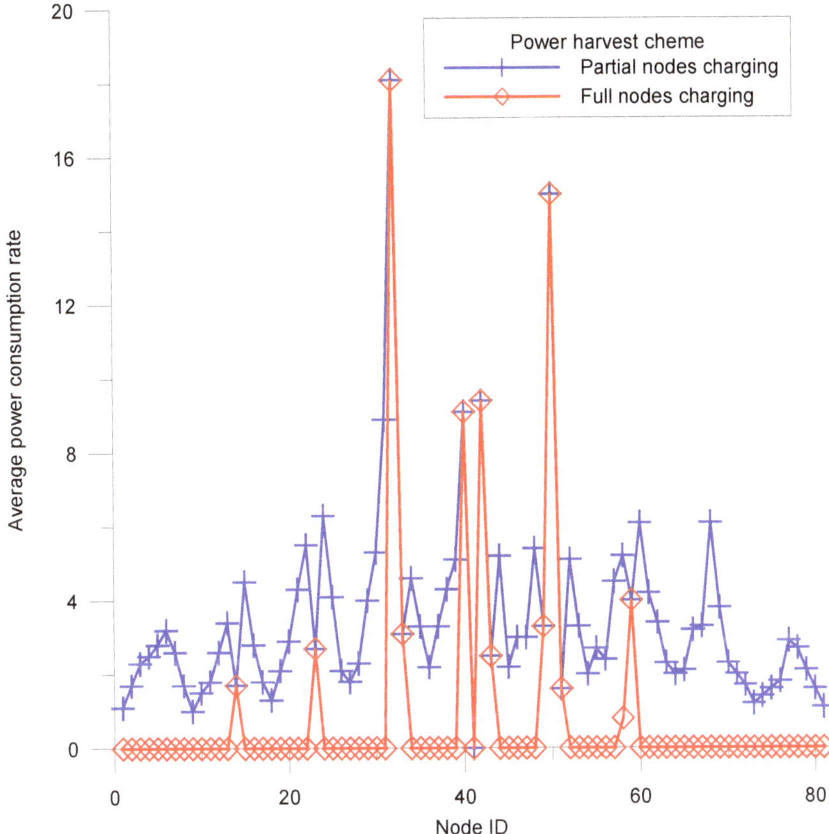

Figure 9. The power consumption rate distribution of the power harvest scheme in SPT-EBR.

For network lifecycle evaluation, the initial energy at each node is set to 0.6 J, and the network lifecycle is evaluated by operation rounds. The energy consumption per transmission bit is 50 nJ/bit and the data is 1000 bits in length [35]. The primary node with the highest power consumption rate is considered as the hotspot, which will limit the overall network lifecycle. Thus, the network lifecycle is inversely proportional to the highest power consumption rate of the hotspot. An additional model, called constant power charging rate is also designed in the energy harvesting scheme to achieve the everlasting WSNs. Figure 10 shows the network lifecycle of both SPT-EBR and Dijkstra algorithms. The power consumption scheme demonstrates an improvement of almost 52% over the Dijkstra scheme, and the power harvest scheme almost doubles the lifecycle than that of the Dijkstra scheme. In addition, the constant power charging rate can extend almost 6.5 times than the Dijkstra scheme in the overall network lifecycle. From the results, it can be concluded that the immortal WSNs can be achieved when the energy harvesting rate is greater than the power consumption rate, especially for the hotspot nodes.

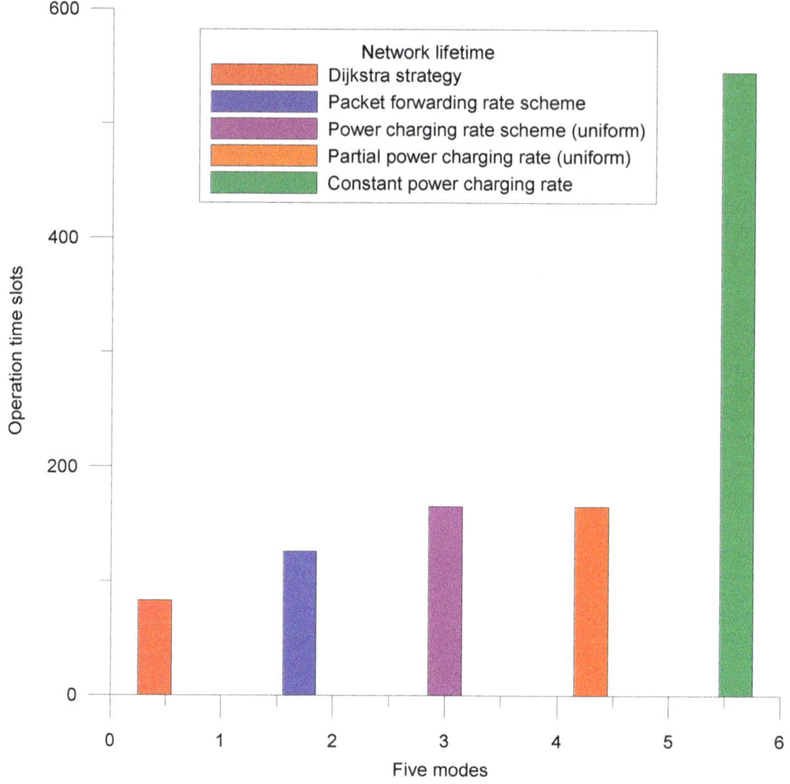

Figure 10. The network lifecycle of SPT-EBR and Dijkstra algorithms.

Figure 11 shows that the battery charging rate Δb of three cases includes the low, the medium and the high charging conditions for the worst hotspot node of packet transmission. The constant charging rate for the low case is 10 units/round, the medium case is 20 units /round, and the high case is 30 units/round. The high battery charging rate case achieves the best energy harvesting performance since the deployment cost is the highest. Because the parameter Δb is greater than 0 in the high charging rate case, it means that the battery of the worst sensing node is always in the positive charging state and will not be affected by the power consumption of packet transmission. In other words, the immortal WSNs can be achieved by control the positive charging rate for the worst hotspot node during packet transmission. In addition, the peak power consumption rate is 36/27/18/9 for the four intra-layers from the sink for the conventional scheme (blue curve) and is 23.2/11.9/6.9/3.9 for the SPT-EBR scheme in Figure 7, respectively. The peak power consumption rate means that the minimum charging power for each layer and thus the proposed method can save the charging power about 55.7% in average in each operation round.

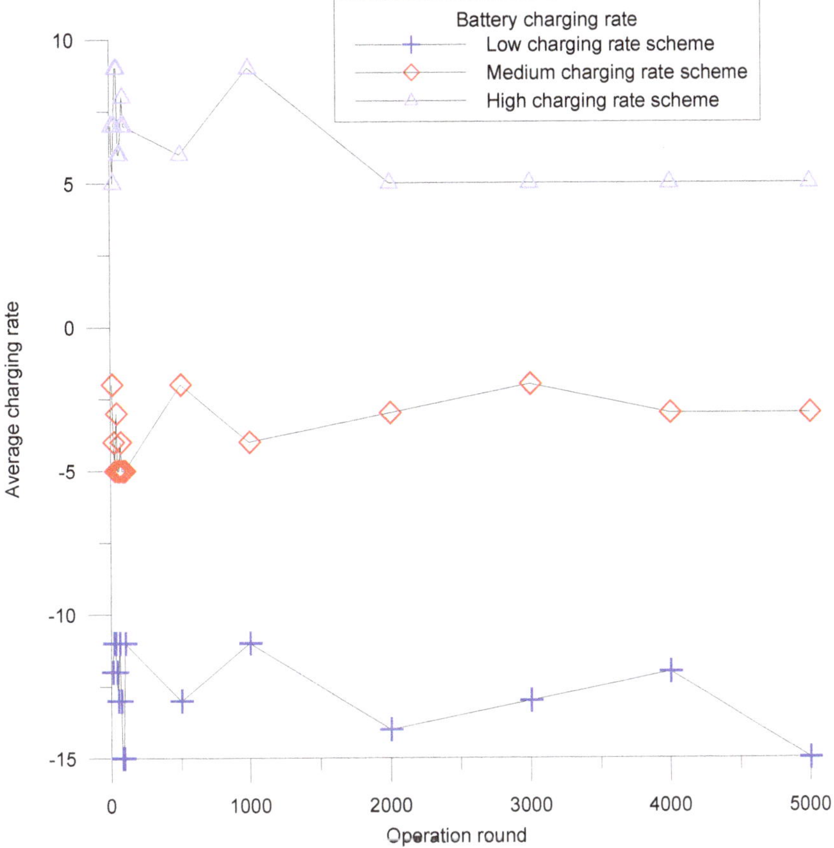

Figure 11. The battery charging rate of the worst hotspot node for SPT-EBR.

Figure 12 shows the routing computation cost of power consumption scheme and energy harvesting scheme in the SPT-EBR and the Dijkstra algorithm. It can be observed that the Dijkstra algorithm achieves the least computation time to make the routing path convergence. To achieve lower power consumption rate and higher network lifecycle, the power consumption scheme spends additional 34% time cost for balanced routing computation. In the energy harvesting scheme, three strategies including the full nodes harvesting, the partial nodes harvesting and the constant harvesting achieve better network lifecycle extension but spends additional 72% time cost to generate the balanced routing paths. From the above results, it is better to spend routing computation cost in the early period rather than the late-remedy strategy in the maintenance phase to extend the network lifecycle.

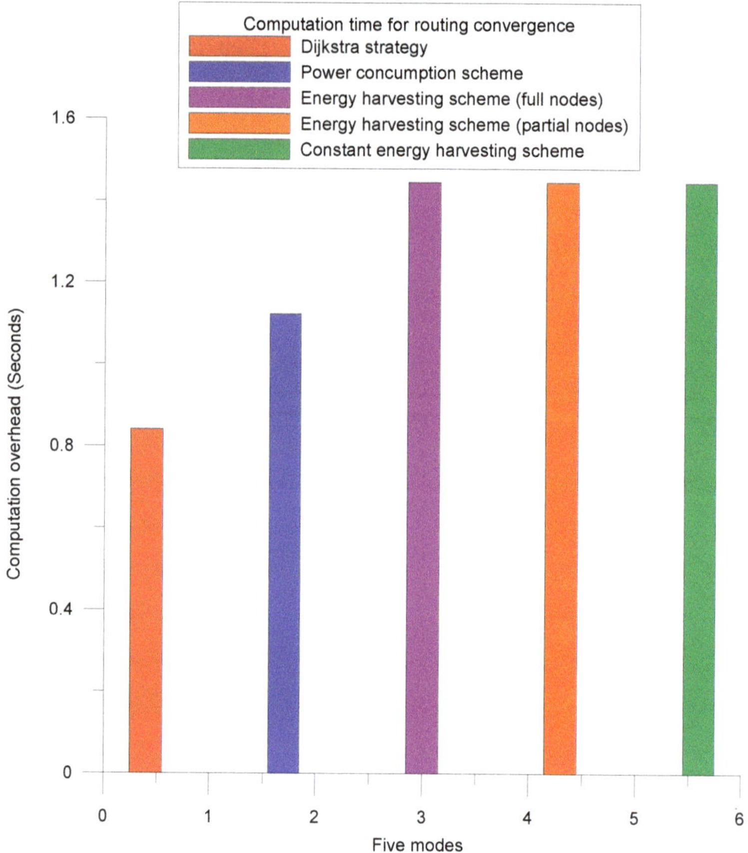

Figure 12. The routing computation cost of SPT-EBR and Dijkstra algorithms.

6. Conclusions

To joint consider the energy balanced routing and energy harvesting benefits, a Shortest Path Tree with Energy Balance Routing strategy (SPT-EBR) is proposed in the article. In SPT-EBR, two methods are presented, including the power consumption and the energy harvesting schemes to select the forwarding node according to the awareness factors of link weight. First, the packet forwarding rate factor was considered in the power consumption scheme to update the link weight for the sensors with higher power consumption and mitigate the traffic load of hotspot nodes to achieve the energy balance network. Then, the energy harvesting scheme combines both the packet forwarding rate and the power charging rate factors together to update the link weight with a new battery charging rate factor for the hotspots. Finally, simulation results validate that both power consumption and energy harvesting schemes in SPT-EBR achieve better energy balance performance and save more charging power than the conventional shortest path protocol and thus improve the overall network lifecycle. Additionally, it can also be inferred that the immortal WSNs has a high probability to be realized when the energy harvesting rate is greater than the power consumption rate especially for the hotspot nodes. As a result, the design aim of immortal WSNs can be achieved, which profits from the effective harvested energy and load balanced routing strategy.

Author Contributions: C.-M.Y. designed the joint balanced routing and energy harvesting strategy with power consumption and energy harvesting schemes including the corresponding algorithms. M.T. helps to investigate the related work and validates the simulation performances. C.-H.C. and C.-Y.H. performed the simulation and analyzed the results.

Funding: This research was funded by the College of Artificial Intelligence, Yango University, China.

Conflicts of Interest: The authors declare no conflict of interest.

References

1. Xu, L.; Collier, R.; O'Hare, G.M.P. A Survey of Clustering Techniques in WSNs and Consideration of the Challenges of Applying Such to 5G IoT Scenarios. *IEEE Internet Things J.* **2017**, *4*, 1229–1249. [CrossRef]
2. Gungor, V.C.; Lu, B.; Hancke, G.P. Opportunities and challenges of wireless sensor networks in smart grid. *IEEE Trans. Ind. Electron.* **2010**, *57*, 3557–3564. [CrossRef]
3. Ruiz-Garcia, L.; Lunadei, L.; Barreiro, P.; Robla, I. A review of wireless sensor technologies and applications in agriculture and food industry: State of the art and current trends. *Sensors* **2009**, *9*, 4728–4750. [CrossRef] [PubMed]
4. Gisbert, J.R.; Palau, C.; Uriarte, M.; Prieto, G.; Palazón, J.A.; Esteve, M.; López, O.; Correas, J.; Lucas-Estañ, M.C.; Giménez, P.; et al. Integrated system for control and monitoring industrial wireless networks for labor risk prevention. *J. Netw. Comput. Appl.* **2014**, *39*, 233–252. [CrossRef]
5. Xin, H.; Liu, X. Energy-Balanced Transmission with Accurate Distances for Strip-Based Wireless Sensor Networks. *IEEE Access* **2017**, *5*, 16193–16204. [CrossRef]
6. Yan, J.; Zhou, M.; Ding, Z. Recent Advances in Energy-Efficient Routing Protocols for WirelessSensor Networks: A Review. *IEEE Access* **2016**, *4*, 5673–5686. [CrossRef]
7. Gungor, V.C.; Hancke, G.P. Industrial wireless sensor networks: Challenges, design principles, and technical approaches. *IEEE Trans. Ind. Electron.* **2009**, *56*, 4258–4265. [CrossRef]
8. Dietrich, I.; Dressler, F. On the lifetime of wireless sensor networks. *ACM Trans. Sens. Netw.* **2009**, *5*, 1–39. [CrossRef]
9. Zhang, D.; Li, G.; Zheng, K.; Ming, X. An energy-balanced routing method based on forward-aware factor for wireless sensor networks. *IEEE Trans. Ind. Inform.* **2014**, *10*, 766–773. [CrossRef]
10. Pantazis, N.A.; Nikolidakis, S.A.; Vergados, D.D. Energy-effcient routing protocols in wireless sensor networks: A survey. *IEEE Commun. Surv. Tutor.* **2013**, *15*, 551–591. [CrossRef]
11. Guo, W.; Zhang, W. A survey on intelligent routing protocols in wireless sensor networks. *J. Netw. Comput. Appl.* **2014**, *38*, 185–201. [CrossRef]
12. Chang, J.-H.; Tassiulas, L. Maximum lifetime routing in wireless sensor networks. *IEEE/ACM Trans. Netw.* **2004**, *12*, 609–619. [CrossRef]
13. Gupta, H.P.; Rao, S.V.; Yadav, A.K.; Dutta, T. Geographic routing in clustered wireless sensor networks among obstacles. *IEEE Sens. J.* **2015**, *15*, 2984–2992. [CrossRef]
14. Amini, N.; Vahdatpour, A.; Xu, W.; Gerla, M.; Sarrafzadeh, M. Cluster size optimization in sensor networks with decentralized cluster-based protocols. *Comput. Commun.* **2012**, *35*, 207–220. [CrossRef]
15. Han, Z.; Wu, J.; Zhang, J.; Liu, L.; Tian, K. A general self-organized tree-based energy-balance routing protocol for wireless sensor network. *IEEE Trans. Nucl. Sci.* **2014**, *61*, 732–740. [CrossRef]
16. Yu, C.M.; Ku, M.L. Joint Hybrid Transmission and Adaptive Routing for Lifetime Extension of WSNs. *IEEE Access* **2018**, *6*, 21658–21667. [CrossRef]
17. Yetgin, H.; Cheung, K.T.K.; El-Hajjar, M.; Hanzo, L. A Survey of Network Lifetime Maximization Techniques in Wireless Sensor Networks. *IEEE Commun. Surv. Tutor.* **2017**, *19*, 828–854. [CrossRef]
18. Liu, X. Routing Protocols Based on Ant Colony Optimization in Wireless Sensor Networks: A Survey. *IEEE Access* **2017**, *5*, 26303–26317. [CrossRef]
19. Ogundile, O.O.; Alfa, A.S. A Survey on an Energy-Efficient and Energy-Balanced Routing Protocol for Wireless Sensor Networks. *Sensors* **2017**, *17*, 1084. [CrossRef]
20. Sudevalayam, S.; Kulkarni, P. Energy harvesting sensor nodes: Survey and implications. *IEEE Commun. Surv. Tutor.* **2011**, *13*, 443–461. [CrossRef]
21. Ulukus, S.; Yener, A.; Erkip, E.; Simeone, O.; Zorzi, M.; Grover, P.; Huang, K. Energy harvesting wireless communications: A review of recent advances. *IEEE J. Sel. Areas Commun.* **2015**, *33*, 360–381. [CrossRef]
22. Lu, X.; Wang, P.; Niyato, D.; Kim, D.I.; Han, Z. Wireless charging technologies: Fundamentals, standards, and network applications. *IEEE Commun. Surv. Tutor.* **2016**, *18*, 1413–1452. [CrossRef]

23. Du, R.; Ozcelikkale, A.; Fischione, C.; Xiao, M. Optimal energy beamforming and data routing for immortal wireless sensor networks. In Proceedings of the 2017 IEEE International Conference on Communications (ICC), Paris, France, 21–25 May 2017.

24. Cheng, D.; Xun, Y.; Zhou, T.; Li, W. An energy aware ant colony algorithm for the routing of wireless sensor networks. In *Intelligent Computing and Information Science (Communications in Computer and Information Science)*; Springer: Berlin, Germany, 2011; Volume 134, pp. 395–401.

25. Luo, J.; Hu, J.; Wu, D.; Li, R. Opportunistic routing algorithm for relay node selection in wireless sensor networks. *IEEE Trans. Ind. Inform.* **2015**, *11*, 112–121. [CrossRef]

26. Kumar, V.; Kumar, S. Energy balanced position-based routing for lifetime maximization of wireless sensor networks. *Ad Hoc Netw.* **2016**, *52*, 117–129. [CrossRef]

27. Kulshrestha, J.; Mishra, M.K. An adaptive energy balanced and energy efficient approach for data gathering in wireless sensor networks. *Ad Hoc Netw.* **2017**, *54*, 130–146. [CrossRef]

28. Liu, L.; Zhang, R.; Chua, K.-C. Multi-antenna wireless powered communication with energy beamforming. *IEEE Trans. Commun.* **2014**, *62*, 4349–4361. [CrossRef]

29. Xu, J.; Zhang, R. A general design framework for MIMO wireless energy transfer with limited feedback. *IEEE Trans. Signal Process.* **2016**, *64*, 2475–2488. [CrossRef]

30. Zeng, Y.; Clerckx, B.; Zhang, R. Communications and signals design for wireless power transmission. *IEEE Trans. Commun.* **2017**, *65*, 2264–2290. [CrossRef]

31. Du, R.; Ozcelikkale, A.; Fischione, C.; Xiao, M. Towards Immortal Wireless Sensor Networks by Optimal Energy Beamforming and Data Routing. *IEEE Trans. Wirel. Commun.* **2018**, *17*, 5338–5352. [CrossRef]

32. Singh, K.; Ku, M.L.; Lin, J.C. Joint power control and energy transfer for energy harvesting relay networks. In Proceedings of the 2016 IEEE International Conference on Communications (ICC), Kuala Lumpur, Malaysia, 22–27 May 2016.

33. Singh, K.; Ku, M.L.; Lin, J.C.; Ratnarajah, T. Toward Optimal Power Control and Transfer for Energy Harvesting Amplify-and-Forward Relay Networks. *IEEE Trans. Wirel. Commun.* **2018**, *17*, 4971–4986. [CrossRef]

34. Gupta, A.; Singh, K.; Sellathurai, M. Time-Switching EH-Based Joint Relay Selection and Resource Allocation Algorithms for Multi-User Multi-Carrier AF Relay Networks. *IEEE Trans. Green Commun. Netw.* **2019**, *3*, 505–522. [CrossRef]

35. Wu, D.; He, J.; Wang, H.; Wang, C.; Wang, R. A hierarchical packet forwarding mechanism for energy harvesting wireless sensor networks. *IEEE Commun. Mag.* **2015**, *53*, 92–98. [CrossRef]

MDPI

St. Alban-Anlage 66

4052 Basel

Switzerland

Tel. +41 61 683 77 34

Fax +41 61 302 89 18

www.mdpi.com

Energies Editorial Office

E-mail: energies@mdpi.com

www.mdpi.com/journal/energies